T0142101

Studies in Systems, Decision and Control

Volume 182

Series editor

Janusz Kacprzyk, Polish Academy of Sciences, Warsaw, Poland
e-mail: kacprzyk@ibspan.waw.pl

The series "Studies in Systems, Decision and Control" (SSDC) covers both new developments and advances, as well as the state of the art, in the various areas of broadly perceived systems, decision making and control–quickly, up to date and with a high quality. The intent is to cover the theory, applications, and perspectives on the state of the art and future developments relevant to systems, decision making, control, complex processes and related areas, as embedded in the fields of engineering, computer science, physics, economics, social and life sciences, as well as the paradigms and methodologies behind them. The series contains monographs, textbooks, lecture notes and edited volumes in systems, decision making and control spanning the areas of Cyber-Physical Systems, Autonomous Systems, Sensor Networks, Control Systems, Energy Systems, Automotive Systems, Biological Systems, Vehicular Networking and Connected Vehicles, Aerospace Systems, Automation, Manufacturing, Smart Grids, Nonlinear Systems, Power Systems, Robotics, Social Systems, Economic Systems and other. Of particular value to both the contributors and the readership are the short publication timeframe and the world-wide distribution and exposure which enable both a wide and rapid dissemination of research output.

More information about this series at http://www.springer.com/series/13304

Irina V. Gashenko · Yulia S. Zima
Armenak V. Davidyan
Editors

Optimization of the Taxation System: Preconditions, Tendencies, and Perspectives

 Springer

Editors
Irina V. Gashenko
Rostov State University of Economics
Rostov-on-Don, Russia

Armenak V. Davidyan
Rostov State University of Economics
Rostov-on-Don, Russia

Yulia S. Zima
Rostov State University of Economics
Rostov-on-Don, Russia

ISSN 2198-4182 ISSN 2198-4190 (electronic)
Studies in Systems, Decision and Control
ISBN 978-3-030-13181-4 ISBN 978-3-030-01514-5 (eBook)
https://doi.org/10.1007/978-3-030-01514-5

© Springer Nature Switzerland AG 2019
Softcover re-print of the Hardcover 1st edition 2019
This work is subject to copyright. All rights are reserved by the Publisher, whether the whole or part
of the material is concerned, specifically the rights of translation, reprinting, reuse of illustrations,
recitation, broadcasting, reproduction on microfilms or in any other physical way, and transmission
or information storage and retrieval, electronic adaptation, computer software, or by similar or dissimilar
methodology now known or hereafter developed.
The use of general descriptive names, registered names, trademarks, service marks, etc. in this
publication does not imply, even in the absence of a specific statement, that such names are exempt from
the relevant protective laws and regulations and therefore free for general use.
The publisher, the authors and the editors are safe to assume that the advice and information in this
book are believed to be true and accurate at the date of publication. Neither the publisher nor the
authors or the editors give a warranty, express or implied, with respect to the material contained herein or
for any errors or omissions that may have been made. The publisher remains neutral with regard to
jurisdictional claims in published maps and institutional affiliations.

This Springer imprint is published by the registered company Springer Nature Switzerland AG
The registered company address is: Gewerbestrasse 11, 6330 Cham, Switzerland

Contents

Introduction

Taxation system is one of the most important elements of the institutional environment for conducting entrepreneurial activities in the economic system. Its effectiveness determines the successfulness of the state budget and development of business, which determines the importance and significance of the problem of taxation optimization. The purpose of the book is to determine the perspectives and to develop the system of measures for reforming the taxation system of modern Russia for its optimization. It provides an authors' opinion on the issue of tax optimization in the context of development of small and medium entrepreneurship and offer innovational mechanisms for fighting tax evasion.

Emphasis on small and medium business in the context of modern Russia is the peculiarity of the book, which distinguishes it from other publications on this topic. This issue was selected for research due to the fact that existing publications in the sphere of taxation are usually devoted to studying general fiscal problems in a state, while in the conditions of new economy, the role of small and medium business grows, and peculiarities and perspectives of taxation optimization should be studied by the modern scientific society.

Another peculiarity of the book, which ensures its topicality and attractiveness for the wide audience is studying the process of optimization of the system of taxation not in the isolated manner but in connection to the tendency of informatization (digitization) of the modern economy. This ensures not only the uniqueness of the approach to treatment of optimality of taxation system but also the originality of the offered authors' recommendations for achieving it in economic practice of modern socioeconomic systems by the example of Russia.

The basic hypothesis, which became the initial point for conducting a complex of scientific and practical research, presented in this book, consists in the idea that the current informatization (digitization) opens new possibilities and perspectives for optimization of the taxation system. The purpose of the book is to study the actual tendencies and preconditions to optimize the taxation system of modern Russia and to substantiate expedience and to develop practical recommendations for using the possibilities of its informatization (digitization) in this process.

The target audience of the book includes postgraduates, lecturers of higher educational establishments, and researchers who study foundations of modern macroeconomics. Based on the authors' conclusions and results, representatives of this target audience can build their current and future scientific studies. The book has seven parts.

The first part is devoted to conceptual foundations of taxation, namely the notion and essence of taxes and taxation, functions of taxes, structural analysis of state's taxation system, and classification of taxes and their characteristics.

The second part views the theory and practice of taxation system's functioning: principles and methods of taxation, essence of tax administration and control and state's tax policy, general and special taxation regimes, analysis of tax load, and process of tax payments optimization.

In the third part, the authors determine the perspective tools of optimization of taxation system: tax holidays are viewed as a perspective tool of tax stimulation of modernization of entrepreneurship; peculiarities of labor taxation are studied in the conditions of formation of social market economy; restructuring of tax obligations is viewed as a perspective direction of diversification of modern Russia's economy; conceptual foundations are determined and methodology of evaluating the effectiveness of state tax policy is given.

The fourth part is devoted to studying the optimization of taxation at the level of separate economic subjects. The authors consider the issues of tax optimization and tax load within the management of taxation at a modern company, analyze the necessity for personal tax management, and study the tax consciousness and formulate the "free rider problem" in taxes.

In the fifth part, the authors study the current state and tendencies of reformation of the modern Russia's taxation system: the methods of optimization of the taxation system with the help of informatization and problem of their application in Russia are studied; ways and methods of simplification of the taxation system with the help of informatization for supporting small and medium business are offers; informatization is viewed as a mechanism of tax evasion fighting.

The sixth part is devoted to substantiating top-priority directions of taxation optimization in modern Russia. The authors offer a model of development and implementation of effective taxation policy in modern Russia, study tax crisis and crisis management in the system of taxation of modern Russia, and digitization of taxes as a top-priority direction of optimization of the taxation system in modern Russia.

In the seventh part, the authors' recommendations for the optimization of taxation in the conditions of information economy are given; the concept of tax stimulation of modern entrepreneurship's informatization is provided; the mechanism of tax stimulation of Industry 4.0 in modern Russia is offered; peculiarities of taxation of Internet companies as the key subjects of the information economy are considered; the model of well-balanced taxation for overcoming the shadow economy in modern Russia is offered; the strategy of provision of tax security of the state in the conditions of information economy is offered.

Part I
Conceptual Foundations of Taxation

The Notion and the Essence of Taxes and Taxation. Functions of Taxes

Irina V. Gashenko, Yuliya S. Zima and Armenak V. Davidyan

Abstract *Purpose* The purpose of the study is to investigate available interpretations of the notion and the essence of taxes and taxation and to represent the comprehensive one, as well as to reveal the relation between competitiveness and successful performance of the tax system. *Methodology* The methodology unit to proof a hypothesis under suggestion is based on the methods of regression and correlation analysis that allows making a model of pair linear regression and calculating the correlation of the world tax systems competitiveness and values of their successful performance. The study is conducted on the data of the end of 2017–the beginning of 2018. To ensure the representativeness of the research outcomes, we use the countries located in different regions of the world with different level and rate of social and economic development as objects of the study. *Results* The analysis showed that regression and correlation dependencies between successful performance of the world tax systems and their competitiveness are low or moderate. Consequently, the relation between them is very weak or absent. Therefore scientifically, it's more logical to define the competitiveness of the tax system not as its target property, but as one of the present system functions; it is expedient to consider their successful performance as the most important goal of this system. It makes possible to overcome the opposition of competitiveness of the tax system and its functions, and also give a fuller description. Therefore, we offer the following functions of the tax system: revenue, regulating, social, supervising, and the one providing the competitiveness of this system. *Recommendations* It is recommended to revise the methodological provisions and approaches to the theory of taxes and taxation and to upgrade empirical ones within the terms of this theory due to its clarified conceptual provisions.

I. V. Gashenko (✉) · Y. S. Zima · A. V. Davidyan
Rostov Institute of National Economy,
Rostov State Economic University, Rostov-on-Don, Russian Federation
e-mail: gaforos@rambler.ru

Y. S. Zima
e-mail: zima.julia.sergeevna@gmail.com

A. V. Davidyan
e-mail: dav_121192@mail.ru

© Springer Nature Switzerland AG 2019
I. V. Gashenko et al. (eds.), *Optimization of the Taxation System: Preconditions, Tendencies, and Perspectives*, Studies in Systems, Decision and Control 182,
https://doi.org/10.1007/978-3-030-01514-5_1

Keywords Taxes · Taxation · Tax system · Competitiveness · Tax functions

JEL Classification E62 · H20 · K34

1 Introduction

Importance and relevance of taxes and taxation for modern economic systems is practically assured and stressed by all scientists and experts in the area of economic government administration. However, the complexity and versatility of taxes and taxation brings an ambiguity in the interpretation of their notion and essence, which presently need clarification in the frames of each scientific research depend on its objectives.

Heterogeneity and incoherence, as well as inconsistency of available definitions of concept and essence of taxes and taxation undermines scientific foundations for studying this economic phenomenon (and process) and causes an issue to systemize a knowledge accumulated in this area and to compose a clarified integral notion and essence of taxes and taxation.

Due to globalization, by now taxes and taxation have become a way to provide a global competitiveness of economic systems instead of the one to solve internal macroeconomic tasks. In this regard, the world tax systems competitiveness is ranked, for example, in accordance with the most established one, International Tax Competitiveness Index, annually represented by Tax Foundation on the values of corresponding index calculated by this international organization.

Performances that determine the tax systems competitiveness of are their attractiveness for international investors (the lower tax rates and the higher stability and predictability of tax policy, the more its attractiveness) and its neutrality (the lesser omissions related to the application of tax benefits, the more neutrality). At the same time, internal needs due to successful performance of functions imposed on them, and the peculiarities of economic systems are not taken into account.

For this reason, in this study we put forward a hypothesis that the competitiveness of the tax system does not guarantee a successful performance of functions (and vice versa). The purpose of the study is to investigate available interpretations of the notion and the essence of taxes and taxation and to represent the comprehensive one, as well as to reveal the relation between competitiveness and successful performance of the tax system.

2 Materials and Method

As follows from the systematization of available scientific knowledge contained in the latest expert studies on taxes and taxation, we build up the following author-owned classification of fundamental approaches to reveal the essence and functions of taxes and taxation (Table 1).

The methodology unit to proof a hypothesis under suggestion is based on the methods of regression and correlation analysis that allows making a model of pair linear regression and calculating the correlation of the world tax systems competitiveness and values of their successful performance. The study is conducted on the data of the end of 2017–the beginning of 2018. To ensure the representativeness of the research outcomes, we use the countries located in different regions of the world with different level and rate of social and economic development as objects of the study.

3 Results

In accordance with Table 1, a revenue approach puts as a priority a revenue function of taxes and taxation concerning assurance of the state budget revenues (when admitting all other functions). The values of successful performance are a budget balance and a country's credit rating contained in the Global Competitiveness Report of the World Economic Forum (Table 2 and Fig. 1).

As can be seen from Fig. 1, with increase in the tax system competitiveness of 1 point, the state budget balance of the countries under study is increased of 0.0192% of the GDP, and the credit rating is reduced of 0.0723 points. Low values of

Table 1 Fundamental approaches to reveal the essence and functions of taxes and taxation

Approach	Key function of taxes and taxation	Essence of taxes and taxation	Representatives of approach
Revenue	Revenue	A way to assure state budget revenues	Aslanov (2015), Miller et al. (2016), Nerudová and Dvořáková (2014), Nerudová and Solilová (2017)
Regulatory	Regulating	A way to administer an economic activity	Musayev and Musayeva (2018), Popkova et al. (2018a, b), Shokeen et al. (2017), Zhang et al. (2016)
Social	Social	A way to ensure a social justice	Dagan (2017), Dallyn (2017), Faizal et al. (2017), Hearson (2017)
Supervision	Supervising	Self-controlled (tax) system	Cui and Notteboom (2017), Spartà and Stabile (2017), Verma et al. (2017), Wu and Lu (2018)

Source Compiled by the authors

Table 2 Original data to analyze the dependence of successful revenue performance of the tax system and its competitiveness

Country	The tax system competitiveness, points from 0 to 100 (position) (x)	State budget balance, % of the GDP (y)	Country's credit rating, points from 0 to 100 (max) (y)
Estonia	100.0 (1)	0.3 (12)	76.8 (24)
New Zealand	88.7 (2)	0.6 (9)	86.3 (15)
Switzerland	85.2 (3)	−12.4 (129)	20.0 (125)
Turkey	73.7 (11)	−2.3 (54)	52.6 (68)
The United Kingdom	70.8 (14)	−3.1 (72)	88.9 (13)
Norway	70.7 (15)	2.9 (4)	94.8 (2)
Canada	69.1 (17)	−1.9 (48)	92.3 (8)
Japan	66.8 (22)	−4.2 (93)	82.0 (19)
Germany	66.6 (23)	0.8 (8)	94.7 (3)
Mexico	62.2 (25)	−2.9 (70)	71.0 (34)
USA	55.1 (30)	−4.4 (95)	93.4 (4)
Italy	47.7 (34)	−2.4 (61)	68.4 (37)
France	43.4 (35)	−3.3 (76)	84.0 (16)
Russia	34.42 (58)	−3.7 (84)	54.2 (61)

Source Compiled by the authors based on: Tax Foundation (2018), World Economic Forum (2018)

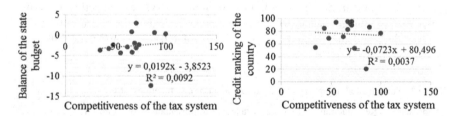

Fig. 1 Regression curves displaying the dependence of successful revenue performance of the tax system on its competitiveness. *Source* Constructed by authors

correlation coefficients in both cases (0.92 and 0.37% respectively) evidence a statistical insignificance of dependencies revealed.

The representatives of regulatory approach consider taxes and taxation as a way to regulate an economic activity (a key function is regulating) through restraining one activity and promoting another one (tax privileges) for increase of business activity, which property is the Index of Economic Freedom composed by The Heritage Foundation, and the growth of living standards of the population described by a corresponding rating Legatum Institute (Table 3 and Fig. 2).

Table 3 Original data to analyze the dependence of successful regulating performance of the tax system and its competitiveness

Country	The tax system competitiveness, points from 0 to 100 (max) (position) (x)	Place in a global rating due to a living standard of the population (y)	Index of Economic Freedom, points from 0 to 100 (max) (position) (y)
Estonia	100.0 (1)	27	78.8 (7)
New Zealand	88.7 (2)	2	84.2 (3)
Switzerland	85.2 (3)	4	55.9 (123)
Turkey	73.7 (11)	88	65.4 (58)
The United Kingdom	70.8 (14)	10	78.0 (8)
Norway	70.7 (15)	1	74.3 (23)
Canada	69.1 (17)	8	77.7 (9)
Japan	66.8 (22)	23	72.3 (30)
Germany	66.6 (23)	11	74.2 (25)
Mexico	62.2 (25)	61	64.8 (63)
USA	55.1 (30)	18	75.7 (18)
Italy	47.7 (34)	30	62.5 (79)
France	43.4 (35)	19	63.9 (71)
Russia	34.42 (58)	101	58.2 (107)

Source Compiled by the authors based on: Tax Foundation (2018), Legatum Institute (2018), The Heritage Foundation (2018)

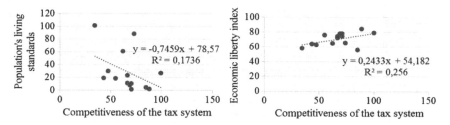

Fig. 2 Regression curves displaying the dependence of successful regulating performance of the tax system on its competitiveness. *Source* Constructed by authors

As can be seen from Fig. 2, with increase in the tax system competitiveness of 1 point, the living standard of the population is reduced of 0.7459 items, and the value of the economic freedom index is increased of 0.2433 points. Moderate values of correlation coefficients in both cases (17.36 and 25.60% respectively) evidence a statistical insignificance of dependencies revealed.

Within a social approach, taxes and taxation are perceived as a way to ensure a social justice with focus on their social function (when admitting all other

Table 4 Original data to analyze the dependence of successful social performance of the tax system and its competitiveness

Country	The tax system competitiveness, points from 0 to 100 (position) (x)	The European index of social justice, points from 0 to 10 (position) (y)
Estonia	100.0 (1)	6.19 (12)
The United Kingdom	70.8 (14)	6.22 (11)
Germany	66.6 (23)	6.71 (7)
Italy	47.7 (34)	4.84 (25)
France	43.4 (35)	6.29 (10)
Denmark	67.0 (21)	7.39 (1)
Sweden	81.8 (6)	7.31 (2)
Finland	68.2 (19)	7.14 (3)
Poland	54.4 (31)	5.79 (15)
Portugal	51.9 (33)	5.36 (20)
Spain	59.8 (28)	4.96 (24)
Greece	57.2 (29)	3.70 (28)

Source Compiled by the authors based on: Tax Foundation (2018), Bartelsmann Stiftung (2018)

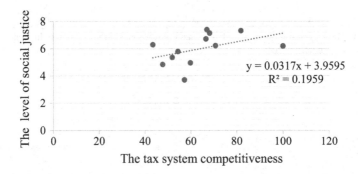

Fig. 3 The regression curve displaying the dependence of successful social performance of the tax system on its competitiveness. *Source* Compiled by the authors

functions). The performance of social justice is the social justice index composed by Bartelsmann Stiftung for the European countries (Table 4 and Fig. 3).

As can be seen from Fig. 3, with increase in the tax system competitiveness of 1 point, the level of social justice is increased of 0.0317 points. The moderate value of the correlation coefficient (19.59%) evidences a statistical insignificance of the dependency revealed.

A supervision approach stresses an importance of a supervising function of taxes and taxation (when admitting all other functions) defining them as a self-regulatory (tax) system. It is believed that the more complex the tax system, the more difficult to exercise its self-supervision. The performance of the tax system complexity is a

Table 5 Original data to analyze the dependence of successful supervising performance of the tax system and its competitiveness

Country	The tax system complexity, points from 0 to 100 (position) (x)	The tax system competitiveness, points from 0 to 100 (position) (y)
Estonia	48.70 (14)	100.0 (1)
New Zealand	34.50 (9)	88.7 (2)
Switzerland	28.80 (19)	85.2 (3)
Turkey	41.10 (88)	73.7 (11)
The United Kingdom	30.70 (23)	70.8 (14)
Norway	37.50 (28)	70.7 (15)
Canada	20.90 (16)	69.1 (17)
Japan	47.40 (68)	66.8 (22)
Germany	48.90 (41)	66.6 (23)
Mexico	52.10 (115)	62.2 (25)
USA	43.80 (36)	55.1 (30)
Italy	48.0 (112)	47.7 (34)
France	62.20 (54)	43.4 (35)
Russia	47.50 (52)	34.42 (58)

Source Compiled by the authors based on: Tax Foundation (2018), World Bank Group, PricewaterhouseCoopers (2018)

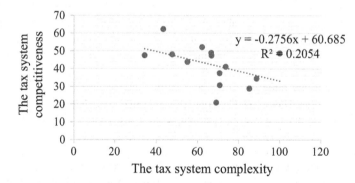

$$y = -0.2756x + 60.685$$
$$R^2 = 0.2054$$

Fig. 4 Regression curves displaying the dependence of the tax system competitiveness on successful supervising performance. *Source* Compiled by the authors

corresponding index calculated by the World Bank Group and Pricewater houseCoopers (Table 5 and Fig. 4).

As can be seen from Fig. 4, with increase in the tax system complexity of 1 point, its competitiveness is reduced of 0.2756 points. The moderate value of the correlation coefficient (20.54%) evidences a statistical insignificance of the dependency revealed.

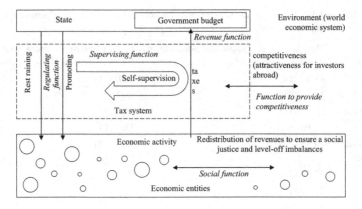

Fig. 5 Complex modern concept of the essence and functions of taxes and taxation *Source* Compiled by the authors

The analysis showed that regression and correlation dependencies between successful performance of the world tax systems and their competitiveness are low or moderate. Consequently, the relation between them is very weak or absent. Therefore scientifically, it's more logical to define the competitiveness of the tax system not as its target property, but as one of the present system functions; it is expedient to consider their successful performance as the most important goal of this system.

It makes possible to overcome the opposition of competitiveness of the tax system and its functions, and also give a fuller description. Accordingly, in this paper we offer the following functions of the tax system: revenue, regulating, social, supervising, and the one providing the competitiveness of this system, which is displayed in Fig. 5.

Figure 5 exhibits integrity (unity and strong interrelation) of the functions of taxes and taxation. Based on the concept suggested, the definition of taxes and taxation in this paper is a way to regulate economic activities aimed at bringing it into compliance with national priorities (performing a regulatory function), to redistribute revenues for ensuring social justice and leveling-off imbalances (social function), providing the state budget replenishment (revenue function) and global competitiveness of the economic system (the function to provide competitiveness) under condition of self-supervision (supervising function).

4 Conclusions

Thus, hypothesis under suggestion has been proofed: multiple tests of nature and strength of relation between the tax systems competitiveness and their successful performances on the statistical data by sampling of different countries of the world have demonstrated the absence of their clear and sustainable dependency. Systematization of scientific interpretations of the notion and the essence of taxes and taxation allowed revealing four functions of taxes and taxation that are recognized by modern scientists: revenue, regulating, social and supervising.

It is also has been established that authors of today mostly prefer only one function without taking into account others, which interferes the integral perception of the tax system. Therefore, we have offered a comprehensive modern concept of the essence of taxes and taxation enabling to combine all their functions and add a new one—providing competitiveness. Clarification of conceptual provisions of the theory of taxes and taxation needs revising the methodological provisions hereof and upgrading approaches to empirical research herein.

References

Aslanov, I. (2015). Government budget revenues in Azerbaijan: The tax burden and the role of the oil factor. *Central Asia and the Caucasus, 16*(3–4), 137–155.

Bartelsmann Stiftung. (2018). *Social justice in the EU—Index report 2017*. URL: https://www. bertelsmann-stiftung.de/en/publications/publication/did/social-justice-in-the-eu-index-report-2017-1/. Data accessed 02.06.2018.

Cui, H., & Notteboom, T. (2017). Modelling emission control taxes in port areas and port privatization levels in port competition and co-operation sub-games. *Transportation Research Part D: Transport and Environment, 56,* 110–128.

Dagan, T. (2017). International tax and global justice. *Theoretical Inquiries in Law, 18*(1), 1–35.

Dallyn, S. (2017). An examination of the political salience of corporate tax avoidance: A case study of the Tax Justice Network. *Accounting Forum, 41*(4), 336–352.

Faizal, S. M., Palil, M. R., Maelah, R., & Ramli, R. (2017). Perception on justice, trust and tax compliance behavior in Malaysia. *Kasetsart Journal of Social Sciences, 38*(3), 226–232.

Hearson, M. (2017). The challenges for developing countries in international tax justice. *Journal of Development Studies, 2*(1), 1–7.

Legatum Institute. (2018). *The Legatum prosperity index 2017*. URL: http://www.prosperity.com/rankings. Data accessed 02.06.2018.

Miller, A. E., Gorlovskaya, I. G., & Miller, A. A. (2016). Development of a mechanism of increasing tax revenues to regional budgets through the development of the securities market. *Journal of Internet Banking and Commerce, 21,* 18–23.

Musayev, A., & Musayeva, A. (2018). A study of the impact of underground economy on integral tax burden in the proportional growth model under uncertainty. *Advances in Fuzzy Systems, 2018,* 6309787.

Nerudová, D., & Dvořáková, V. (2014). The estimation of financial transaction tax revenues as a new own resource of European Union Budget. *Ekonomicky casopis, 62*(9), 945–958.

Nerudová, D., & Solilová, V. (2017). Common consolidated corporate tax base system re-launching: Simulation of the impact on the slovak budget revenues. *Ekonomicky casopis, 65*(6), 559–578.

Popkova, E. G., Bogoviz, A. V., Lobova, S. V., & Romanova, T. F. (2018a). The essence of the processes of economic growth of socio-economic systems. *Studies in Systems, Decision and Control, 135,* 123–130.

Popkova, E. G., Bogoviz, A. V., Ragulina, Y. V., & Alekseev, A. N. (2018b). Perspective model of activation of economic growth in modern Russia. *Studies in Systems, Decision and Control, 135,* 171–177.

Shokeen, S., Banwari, V., & Singh, P. (2017). Impact of goods and services tax bill on the Indian economy. *Indian Journal of Finance, 11*(7), 65–78.

Spartà, G. T., & Stabile, G. (2017). Tax compliance with uncertain income: a stochastic control model. *Annals of Operations Research, 2*(1), 1–13.

Tax Foundation. (2018). *2017 International tax competitiveness index.* URL: https://taxfoundation.org/2017-international-tax-competitiveness-index/. Data accessed 02.06.2018.

The Heritage Foundation. (2018). *2018 Index of economic freedom.* URL: https://www.heritage.org/index/ranking. Data accessed: 02.06.2018.

Verma, P., Agarwal, S., Kachroo, P., & Krishen, A. (2017). Declining transportation funding and need for analytical solutions: dynamics and control of VMT tax. *Journal of Marketing Analytics, 2*(1), 1–10.

World Bank Group, PricewaterhouseCoopers. (2018). *Paying taxes 2018.* URL: https://www.pwc.com/gx/en/services/tax/publications/paying-taxes-2018/overall-ranking-and-data-tables.html. Data accessed: 02.06.2018.

World Economic Forum. (2018). *The global competitiveness report 2017–2018.* URL: http://www3.weforum.org/docs/GCR2017-2018/05FullReport/TheGlobalCompetitivenessReport2017–2018.pdf. Data accessed 02.06.2018.

Wu, Z., & Lu, X. (2018). The effect of transfer pricing strategies on optimal control policies for a tax-efficient supply chain. *Omega (United Kingdom), 2*(1), 34–39.

Zhang, Y.-Y., Wang, Y.-G., & Tian, Y.-J. (2016). Impact analysis of carbon tax on the economy of modern coal chemical products in China. *Xiandai Huagong/Modern Chemical Industry, 36*(12), 1–4.

Tax System of a State: Federal, Regional and Local Taxes and Fees

Irina V. Gashenko, Yuliya S. Zima and Armenak V. Davidyan

Abstract *Purpose* The purpose of the paper is studying the peculiarities to adjust contradictions in the Russian tax system and drawing up a practical model to organize tax system of the country and to implement a fiscal federalism in modern Russia. *Methodology* The methodology for testing a hypothesis under suggestion includes methods of economic statistics, namely, methods of structural and horizontal analysis, as well as ones of comparative and plan-fact analysis and formalization (graphical representation of data). Information and analytical background of the study is the official statistical data of the Federal Tax Service of the Russian Federation, the Ministry of Finance of the Russian Federation, and the Federal State Statistics Service of the Russian Federation (Rosstat) for 2008–2017. *Results* We revealed deviations from preferable (target) peculiarities to adjust contradictions in the tax system of modern Russia that lead to imbalance of local fiscal subsystems, escalation of deficit in all level-budgets, as well as establishing the dependence of territories on intergovernmental transfers. We have built a practical model for organization of the tax system and for implementation of fiscal federalism in the Russian Federation, which has not only advantages (such as integrity, etc.), but also disadvantages (enhanced limitation of fiscal initiatives and opportunities of territories, multiple overlapping of tax revenue flows, etc.). *Recommendations* We advise an improvement of the tax system in modern Russia through adjustment of revealed disadvantages.

Keywords Tax system of the country · Federal · Regional and local taxes and fees Fiscal federalism

I. V. Gashenko (✉) · Y. S. Zima · A. V. Davidyan
Rostov Institute of National Economy, Rostov State Economic University,
Rostov-on-Don, Russian Federation
e-mail: gaforos@rambler.ru

Y. S. Zima
e-mail: zima.julia.sergeevna@gmail.com

A. V. Davidyan
e-mail: dav_121192@mail.ru

© Springer Nature Switzerland AG 2019

I. V. Gashenko et al. (eds.), *Optimization of the Taxation System: Preconditions, Tendencies, and Perspectives*, Studies in Systems, Decision and Control 182, https://doi.org/10.1007/978-3-030-01514-5_2

JEL Classification E62 · H20 · K34

1 Introduction

Functions of taxes and taxation are performed through establishment, operation and development of the tax system of the country. The fundamental principle of tax systems in most modern countries of the world is a fiscal federalism, which involves a simultaneous delimitation and strong interrelation of federal, regional and local fiscal systems. Following this principle causes three inevitable contradictions in the tax system:

- The collision of federal–national (more internationally-oriented) and territorial–regional and local (more nationally-oriented) interests;
- The collision of interests in public performance (resulting in growth of expenditures and a budget deficit) and in maintaining remunerativeness (with possible surplus) of government budgets;
- The collision between national interests of balanced territorial development (granting temporary federal support to the territories to activate their own economic growth) and territorial interests to rely on federal support (establishing dependence on federal support without following-up their own initiatives to activate economic growth).

The peculiarities of adjusting these contradictions define specific nature of practical models to organize the tax system of the country and to implement fiscal federalism. The Concept of long-term social and economic development of the Russian Federation until 2020 approved by the decree No. 1662-p. as of November 17, 2008 states the following preferred (target) peculiarities of adjusting these contradictions (Government of the Russian Federation 2018a):

- prevailing of territorial interests (as evidenced by declared decentralization of the fiscal system and its commitment to provide regional economy development);
- prevailing of interests in public performance (as evidenced by declared large-scale goals of social and economic development along with a refusal to increase the total tax burden in economy);
- prevailing of national interests in balanced territorial development (as evidenced by declared heightened commitment of the tax system to the issues of levelling up income and enhancing spur impact of the tax system on economic development).

Given that the above-noted concept was developed and adopted before the crisis of the Russian economy in 2009, current hypothesis of this research consists in the author's suggestion that the preferred (target) peculiarities to adjust contradictions of the tax system are not fully achieved (mispresented) in modern Russia due to crisis impact (among other). The purpose of the paper is studying the peculiarities to adjust contradictions in the Russian tax system and drawing up a practical model to organize tax system of the country and to implement a fiscal federalism in modern Russia.

2 Materials and Method

A review of available studies and papers on the topic allowed identifying two basic theoretical models of the state tax system organization and the implementation of fiscal federalism—decentralized and centralized—upon which practical models of countries are to be built.

A decentralized model is applied in countries such as the Great Britain, the USA, Canada, Japan, etc. It involves a high degree of financial (fiscal) independence of territories within the state, wide opportunities to introduce their own taxes, define performances (for example, tax rates), and to make loans, as well as the lack of activity in taking measures aimed at levelling-up imbalances in local budgets and overcoming unremunerativeness. This model is studied in papers by authors such as Del Bo (2018), Haddow (2018), Kimura (2017), Nicholson-Crotty et al. (2006), and Xing (2018).

A centralized model is much more widespread and applied in countries such as Germany, France, Mexico, China, etc. It is related to a low degree of financial (fiscal) independence of territories within the state, their strictly limited opportunities to introduce their own taxes and define performances (for example, tax rates), and to implement loans, as well as activity in taking measures aimed at levelling-up imbalances in local budgets and overcoming unremunerativeness. It is considered in papers by scientists such as Grabova et al. (2018), Gunlicks (2012), Hoyt (2017), and Sun et al. (2017).

Modern Russia follows a centralized theoretical model enshrined in the Tax Code of the Russian Federation (TC RF) N 146-ФЗ as of July 31, 1998. Pursuant to it, the tax system in modern Russia is a three-tiered one with the following categories and corresponding taxes and fees (Government of the Russian Federation 2018b):

- Federal (macro-level category): value added tax, excise duties, personal income tax, corporate income tax, mineral extraction tax, water tax, levies for the use of fauna and aquatic, state duty;
- Regional (meso-level category): corporate property tax, gambling tax, transport tax;
- Local (micro-level category): land tax, individual property tax, sales tax.

The Russian practice of organizing tax system and implementing fiscal federalism is discussed in publications by experts such as Dyukina (2017), Gashenko et al. (2018), Ibragimov et al. (2014), Popkova et al. (2018a, b), Shvetsov and Balikoev (2017).

The methodology for testing a hypothesis under suggestion includes methods of economic statistics, namely, methods of structural and horizontal analysis, as well as ones of comparative and plan-fact analysis and formalization (graphical representation of data). Information and analytical background of the study is the official statistical data of the Federal Tax Service of the Russian Federation, the Ministry of Finance of the Russian Federation, and the Federal State Statistics Service of the Russian Federation (Rosstat) for 2008–2017.

3   Results

To reveal the peculiarities of ratio of federal–national (more internationally-oriented) and territorial–regional and local (more nationally-oriented) interests in modern Russia, let's address the structure of consolidated budget in the Russian Federation (Fig. 1).

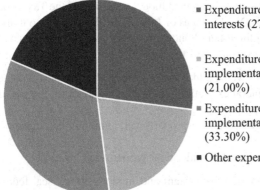

Fig. 1 Structure of revenues and expenditures of the consolidated budget in the Russian Federation in 2017, bln. rub. *Source* Compiled by the authors based on: Federal Tax Service of the Russian Federation (2018), Ministry of Finance of the Russian Federation (2018), Rosstat (2018)

As can be seen from Fig. 1, in the revenue structure of the consolidated budget of the Russian Federation (RUB 17,343.5 billion) proportions of the federal budget (RUB 9162 billion, 52.83%) and consolidated budgets of territories (entities) (RUB 8181.5 billion, 41.17%) are distributed rather equally. In the structure of expenditures, ones related to implementation of national interests are prevailing (RUB 4682.9 billion, 27%), including:

- Expenditures on national issues (RUB 648.6 billion);
- Expenditures on national defense (RUB 1591.5 billion);
- Expenditures on national security and law enforcement activity (RUB 1070.2 billion);
- Expenditures on national economy (RUB 1372.7 billion).

The proportion of federal budget expenditures on national interests' implementation is 21% (RUB 3642.2 billion), including:

- Expenses for housing and public services (RUB 66.7 billion);
- Expenses for social and cultural activities (RUB 2935.5 billion);
- Expenses for education (RUB 343.2 billion);
- Expenses for healthcare (RUB 245.4 billion);
- Expenses for environmental protection (RUB 51.6 billion).

The proportion of local budget expenditures on national interests' implementation is 33.30% (RUB 5775.3 billion), including:

- Expenses for housing and public services (RUB 770.9 billion);
- Expenses for social and cultural activities (RUB 5004.5 billion).

To reveal the peculiarities of ratio between interests in public performance and in maintaining remunerativeness (with possible surplus) of government budgets in modern Russia, let's turn to the balance dynamics in consolidated budget of the Russian Federation in 2008–2017 (Fig. 2).

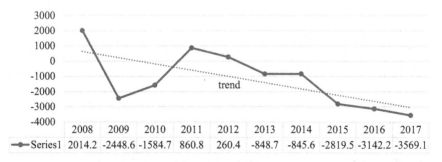

	2008	2009	2010	2011	2012	2013	2014	2015	2016	2017
Series1	2014.2	-2448.6	-1584.7	860.8	260.4	-848.7	-845.6	-2819.5	-3142.2	-3569.1

Fig. 2 Balance dynamics in consolidated budget of the Russian Federation in 2008–2017, RUB billion. *Source* Compiled by the authors based on: Federal Tax Service of the Russian Federation (2018), Rosstat (2018)

As can be seen from Fig. 2, the expenditures of the Russian Federation consolidated budget exceed revenues over almost all period under study, and the balance of this budget is featured by negative dynamics and a downward trend, which evidences an escalation of the budget deficit. In comparison with 2008, when the budget surplus amounted to RUB 2014.2 billion, in 2017 there was a budget deficit of RUB 3569.1 billion. This evidences the prevailing of interests in public performance over the ones in maintaining budget remunerativeness.

To reveal the peculiarities of ratio of the national interests of balanced territorial development (granting temporary federal support to the territories to activate their own economic growth) and the territorial interests to rely on federal support (establishing dependence on federal support without following-up their own initiatives to activate economic growth) in modern Russia, let's turn to balance dynamics in consolidated budgets of the Russian Federation territories (entities) in 2008–2017 (Fig. 3).

As can be seen from Fig. 3, over the entire period under study we observe a negative balance in consolidated budgets of the Russian Federation territories (entities). The highest gravity of their deficit had occurred in 2013 (641.5 RUB billion.). Despite the upward trend, in 2017 still there is a negative balance in consolidated budgets of the Russian Federation territories (entities) in the amount of RUB 24.5 billion. It points to the establishment of the dependence of modern Russia territories on federal support and the prevailing of territorial interests.

Based on the results of our analysis, we built up the following practical model of the tax system organization and the implementation of fiscal federalism in modern Russia (Fig. 4).

As can be seen from Fig. 4, the practical model of the tax system organization and the implementation of fiscal federalism has a complicated structure. In addition to the advantages connected with its high degree of unification, strong interrelation and interdependence of all-level fiscal system, and their mutual support for achieving balance and remunerativeness, this model is featured by the following disadvantages:

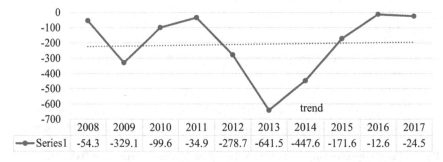

	2008	2009	2010	2011	2012	2013	2014	2015	2016	2017
Series1	-54.3	-329.1	-99.6	-34.9	-278.7	-641.5	-447.6	-171.6	-12.6	-24.5

Fig. 3 Balance dynamics in consolidated budgets of the Russian Federation territories (entities) in 2008–2017, RUB billion. *Source* Compiled by the authors based on: Federal Tax Service of the Russian Federation (2018), Rosstat (2018)

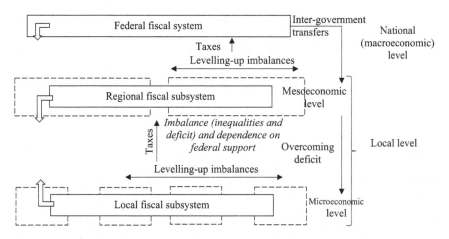

Fig. 4 Practical model of the tax system organization and the implementation of fiscal federalism in Russia. *Source* Compiled by the authors

- incomplete achievement (distortion) of preferred (target) peculiarities to adjust contradictions of the tax system and occurrence of stable unexpected ones related to horizontal imbalances in local fiscal subsystems, deficit and dependence on federal support;
- excessively limited opportunities of the territories in obtaining tax revenues reduced to collection of property taxes, which causes their low (or zero) interest in the development of entrepreneurship and increasing income of the population due to impossible acquisition of revenue from profit tax and personal income tax (since revenues from these taxes are directed to the federal budget);
- oversaturation with tax revenue flows, which are firstly collected in local budgets, then directed to the federal budget and thereafter again are redistributed between the territories.

4 Conclusion

Thus, hypothesis under suggestion is proofed and deviations from the preferable (target) peculiarities to adjust contradictions of the tax system in modern Russia leading to imbalance in local fiscal systems, escalation of budget deficit at all levels, and establishing the dependence of territories on inter-government transfers are revealed.

A practical model built by us to organize the tax system and to implement fiscal federalism in the Russian Federation has not only advantages (such as integrity, etc.), but also disadvantages (enhanced limitation of fiscal initiatives and opportunities of territories, multiple overlapping of tax revenue flows, etc.) It evidences the necessity and the availability of opportunities and prospects for upgrading the tax system in modern Russia.

References

Del Bo, C. F. (2018). Fiscal autonomy and EU structural funds: The case of the Italian regional income tax system. *Public Finance Review, 46*(1), 58–82.

Dyukina, T. O. (2017). Methodological aspects of improvement of tax system in Russia. In *Proceedings of the 29th International Business Information Management Association Conference—Education Excellence and Innovation Management through Vision 2020: From Regional Development Sustainability to Global Economic Growth* (pp. 2811–2819).

Federal Tax Service of the Russian Federation. (2018). *Tax analytics: the structure of revenues in the consolidated budget of the Russian Federation*. URL: https://analytic.nalog.ru/portal/index.ru-RU.htm. Data accessed: 06/03/2018.

Federal Service of State Statistics of the Russian Federation (Rosstat). (2018). *Regions of Russia. Social and economic performances*. URL: http://www.gks.ru/wps/wcm/connect/rosstat_main/rosstat/en/statistics/publications/catalog/doc_1138623506156. Data accessed: 06/03/2018.

Gashenko, I. V., Zima, Y. S., Stroiteleva, V. A., & Shiryaeva, N. M. (2018). The mechanism of optimization of the tax administration system with the help of the new information and communication technologies. *Advances in Intelligent Systems and Computing, 622*, 291–297.

Government of the Russian Federation. (2018a). *The concept of long-term social and economic development of the Russian Federation until 2020 approved by the decree No. 1662-p. as of November 17, 2008*. URL: http://docs.cntd.ru/document/902130343. Data accessed: 06/03/2018.

Government of the Russian Federation. (2018b). *Tax Code of the Russian Federation (TC RF) N 146-ФЗ as of July 31, 1998*. URL: http://www.consultant.ru/document/cons_doc_LAW_19671/. Data accessed: 06/03/2018.

Grabova, O. N., Suglobov, A. E., & Karpovich, O. G. (2018). Evolutionary institutional analysis and prospects of developing tax systems. *Espacios, 39*(16), 21–28.

Gunlicks, A. B. (2012). Legislative competences, budgetary constraints, and the reform of federalism in Germany from the top down and the bottom up. In *Constitutional dynamics in federal systems: Sub-national perspectives* (pp. 61–87). McGill-Queen's University Press.

Haddow, R. (2018). Are Canadian provincial tax systems becoming more regressive? If so, in what respects and why? *Canadian Public Policy, 44*(1), 25–40.

Hoyt, W. H. (2017). The assignment and division of the tax base in a system of hierarchical governments. *International Tax and Public Finance, 24*(4), 678–704.

Ibragimov, M. Y., Ibragimov, R. M., Sinnikova, Y. M., & Tufetulov, A. M. (2014). About the progressive tax system of labor remuneration in Russia. *Asian Social Science, 10*(23), 28–35.

Kimura, S. (2017). Japanese local tax system and decentralization. In *Decentralization and development of Sri Lanka within a unitary state* (pp. 329–364). Singapore: Springer.

Ministry of Finance of the Russian Federation. (2018). *Annual data on the federal budget implementation*.URL: https://www.minfin.ru/en/statistics/fedbud/?id_65=80041&page_id=3847&popup=Y&area_id=65. Data accessed: 06/03/2018.

Nicholson-Crotty, S., Theobald, N. A., & Wood, B. D. (2006). Fiscal Federalism and budgetary tradeoffs in the American States. *Political Research Quarterly, 59*(2), 313–321.

Popkova, E. G., Bogoviz, A. V., Lobova, S. V., & Romanova, T. F. (2018a). The essence of the processes of economic growth of socio-economic systems. *Studies in Systems, Decision and Control, 135*, 123–130.

Popkova, E. G., Bogoviz, A. V., Ragulina, Y. V., & Alekseev, A. N. (2018b). Perspective model of activation of economic growth in modern Russia. *Studies in Systems, Decision and Control, 135,* 171–177.

Shvetsov, Y. G., & Balikoev, V. Z. (2017). Russian tax system: System deficiencies and the direction of the improvement. *Journal of Advanced Research in Law and Economics, 8*(4), 1332–1339.

Sun, Z., Chang, C.-P., & Hao, Y. (2017). Fiscal decentralization and China's provincial economic growth: A panel data analysis for China's tax sharing system. *Quality & Quantity, 51*(5), 1–23.

Xing, J. (2018). Territorial tax system reform and multinationals' foreign cash holdings: New evidence from Japan. *Journal of Corporate Finance, 49,* 252–282.

Classification of Taxes and Their Features

Irina V. Gashenko, Yuliya S. Zima and Armenak V. Davidyan

Abstract *Purpose* The purpose of the paper is to study classifications of taxes and their features, as well as to give macroeconomic description of the tax burden in modern Russia and to evaluate its suitability. *Methodology* The methodological unit of this study includes the method of classification, the method of comparative analysis and a complex of general scientific methods (induction, deduction, synthesis, formalization, etc.). Also, we employed a complex of methods to define the tax burden (from the standpoint of macroeconomics). *Results* The author suggested a new classification of taxes upon criterion of complexity of tax evasion and tax administration, which makes it possible to reveal a risk-related component of the tax system, thereby to study its sustainability. We made a conclusion that the tax burden in modern Russia is unreasonably high in comparison with the level that is acceptable for economic entities, despite lower tax rates against other countries (as illustrated by Germany). Unsuitability of the tax burden with a high probability may be a constraint in overcoming crisis impact on the Russian economy. We have made a macroeconomic description of the tax burden in modern Russia, which allowed revealing both its advantages (sustainability, economic viability for the state, assistance in raising of living standards, transparency, commitment to tax liability and reducing the burden on tax authorities), and its disadvantages (internal orientation, low efficiency, high risk-related component, insufficient approval of tax burden by economic entities). *Recommendations* Overcoming the above-mentioned disadvantages of the tax burden gives prospects to improve the Russian tax system upon classifications proposed.

Keywords Classification of taxes · Tax burden · Modern Russia

I. V. Gashenko (✉) · Y. S. Zima · A. V. Davidyan
Rostov Institute of National Economy, Rostov State Economic University,
Rostov-on-Don, Russian Federation
e-mail: gaforos@rambler.ru

Y. S. Zima
e-mail: zima.julia.sergeevna@gmail.com

A. V. Davidyan
e-mail: dav_121192@mail.ru

© Springer Nature Switzerland AG 2019   23
I. V. Gashenko et al. (eds.), *Optimization of the Taxation System: Preconditions,*
Tendencies, and Perspectives, Studies in Systems, Decision and Control 182,
https://doi.org/10.1007/978-3-030-01514-5_3

JEL Classification E62 · H20 · K34

1 Introduction

Taxes represent a conditional agreement between the state and economic entities on fulfillment of mutual obligations. According to it, economic entities (taxpayers) undertake to pay taxes (in full extent and in due time), and the state—to fulfill functions imposed on it (providing social safeguards and public benefits, etc.).

The tax agreement is characterized by two difficulties. The first is the collision of Parties' interests. Thus, the state is interested in maximizing its tax revenues and minimizing its obligations to economic entities, which in their turn seek to limit their expenditures hereunder and earn the most possible benefits from execution hereof.

The second difficulty is related to the macroeconomic nature of agreement. A great number and diversity of economic entities to which it is a party, causes a collision of their interests, as well as the necessity of joint efforts to assert their rights hereunder. Therefore, to ensure compliance with the tax contract's terms in the economic system, conclusion party (the state) takes into account the peculiarities of this system and selects the most suitable taxes; multiple tax classifications (varieties) provide wide opportunities for this.

Therewith, a topical scientific and practical issue of today is the definition of suitable structural composition of the tax system and features of taxes for each social and economic system due to tax classification. At the same time, one of the most important performances is the tax burden showing the aggregated tax burden on economic entities and their perception of it.

With respect to features of unsuitability of the tax system in modern Russia revealed in the previous chapters and unremunerativeness of all-level budgets, we put forward a hypothesis that the tax burden in the Russian economy is unreasonably high. The purpose of the paper is to study classifications of taxes and their features, as well as to give macroeconomic description of the tax burden in modern Russia and to evaluate its suitability.

2 Materials and Method

The methodological unit of this study includes the method of classification, the method of comparative analysis and a complex of general scientific methods (induction, deduction, synthesis, formalization, etc.). A theoretical review on the given topic has shown that numerous studies of modern scientists and experts are devoted to tax classifications, among which Andrejovská and Hudáková (2016), Bauman and Shaw (2016), Duan et al. (2018), Fang et al. (2017), Garcia and von

Haldenwang (2016), Gashenko et al. (2018), He (2017), Heim (2016), Kim (2018), Popkova et al. (2018a, b).

However, despite a high coverage of issue in present study, it is marked by inconsistency and incompleteness of accumulated knowledge. This publication is dedicated to systematization and expansion of available tax classifications. Also, we revealed three methods to define the tax burden (from the standpoint of macroeconomics):

- method to define the rates of underlying taxes and their comparison with the alternative—a similar economic system (as illustrated by Germany) because it provided a prototype for a centralized theoretical model of the tax system, the most popular in the world and implemented in modern Russia;
- method to define the ratio of tax levies in the consolidated government budget to the state's GDP;
- method to define the shadow economy scale as an index of tax evasion in the economic system.

To make the most complete and reliable macroeconomic description of the Russian tax burden in this paper, we applied the above methods in their entirety.

3 Results

On the back of studying available tax classifications and addition of proprietary ones, we obtained the following results (Table 1).

In Table 1 the first four classifications are the most widespread, the next three classifications are extremely rare in papers by modern authors and elaborated by us for the purposes of this study, and the last classification is proprietary. Thus, most researchers classify taxes by the level of their establishment, subject of taxation, method of levying and the tax treatment.

In some situations, when classifying taxes, we also take into account criteria such as the method of tax assessment and frequency of payment, the method of initiating and calculating taxes, as well as sources of tax return. In our opinion, available classifications should be extended by allocation of taxes due to difficulty of tax evasion and tax administration, which allows revealing a risk-related component of the tax system, thereby studying its sustainability.

Based on the tax classifications reviewed, we will evaluate the tax burden in modern Russia. A comparative analysis of rates of underlying taxes that are typical for centralized theoretical model of the tax system in Russia and Germany in 2018 is displayed in Table 2.

As can be seen from Table 2, the total tax burden on individual economic entities in Germany (58.35%) is much higher than in Russia (43%) with a similar tax burden on corporate economic entities (20.5 and 20% respectively) and a similar level of mixed taxes (VAT: 19% and 18% respectively). However, in Germany, tax

Table 1 Classifications of taxes

Criterion of classification	Types of taxes due to given criterion
Tax level	Federal and territorial (regional and local) taxes
Taxpayer (subject of taxation)	Individual (income tax, personal property tax, etc.), corporate (income tax, corporate property tax, etc.) and mixed (value added tax, transport tax, etc.) taxes
Method of levying taxes	Direct (income tax, profit tax, property taxes, etc.) and direct (value added tax, excise duties, etc.) taxes
Tat treatment	Basic (standard conditions of taxation) and special (preferential conditions of taxation)
Method of tax assessment and frequency of payment	Taxes assessed upon termination of economic transaction and paid in a lump sum (duties, etc.), taxes assessed and paid regularly (property taxes, etc.)
Method of initiation and calculation of taxes	Taxes initiated (through filing a return) and calculated by taxpayers (value added tax, etc.), taxes initiated (through tax assessment notice) and calculated by tax authorities (property taxes)
Sources of tax return	Taxes on internal (value added tax on domestic goods, excise duties on domestic goods, etc.) and external (value added tax on import goods, excise duties on import goods, customs duties, etc.) economic activity
Difficulty of tax evasion and tax administration	Taxes that are featured with difficult evasion and, respectively, simple tax administration (property taxes, etc.) and taxes that are featured with simple evasion and difficult tax administration (personal income tax, value added tax, etc.)

Source Drawn up by authors

Table 2 Comparative analysis of underlying tax rates of Russia and Germany in 2018

Category of taxes	Tax	Rate in Germany (%)	Rate in Russia (%)
Personal taxes	Income tax	14 (up to 45)	13
	Social contributions	38.85	60
	Solidarity tax	5.5	–
	Total	58.35	43
Corporate taxes	Profit tax	15	20
	Solidarity tax	5.5	
	Total	20.5	20
Mixed taxes	Value added tax	19	18
	Total	19	18

Source Drawn up by authors on the data (World Bank 2018a, b)

levies make 45.2% of the GDP (World Bank 2018a), and in Russia—32.8% of the GDP (World Bank 2018b). Therewith, the share of the shadow economy in Germany is 11.97%, and in Russia—38.48% (the International Monetary Fund 2018).

This evidences that the tax burden is negatively adopted by economic entities in Russia amid tolerant attitude in Germany. It should be noted that the reason for this phenomenon may also be a higher development of special tax treatments. The revenue structure of consolidated budget in the Russian Federation in 2017 is presented in Fig. 1.

As can be seen from Fig. 1, the biggest share in the revenues structure of the Russian consolidated budget is taken by mineral extraction taxes (25%). It is followed by income taxes—profit tax (20%) and personal income tax (20%), then value added tax (19%), excise duties (9%) and property taxes (7%). The highest arrears are related to value added tax and income tax. On the above-stated, the macroeconomic description of the tax burden in modern Russia are made by us and presented in Fig. 2.

As can be seen from Fig. 2, the basis of the tax burden in modern Russia is a mineral extraction tax with low tax rates providing the maximum revenue for the state at a minimum risk. The core of the tax burden is income taxes with high risk at a high income, and value added tax with an average risk at average income; in both cases we deal with average (in comparison with other countries) tax rates.

The tax burden is supplemented by excise duties and property taxes with high tax rates and a minimum income at a minimum risk. As a result, we achieve the average replenishment of the government budget in the event of side effects

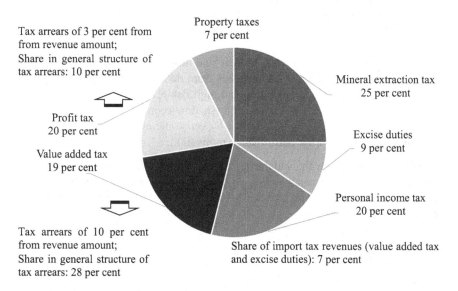

Fig. 1 Revenue structure of the Russian consolidated budget in 2017. *Source* Drawn by authors on the data (Federal Tax Service of the Russian Federation 2018; Rosstat 2018)

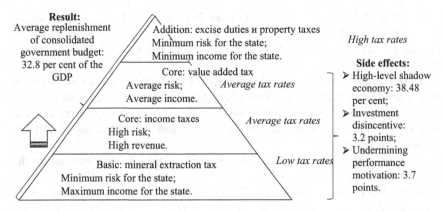

Fig. 2 Macroeconomic description of the tax burden in modern Russia. *Source* Drawn by authors

associated with a high-level shadow economy, investment disincentive and undermining performance motivation.

In connection with noted tax classifications, the tax burden in modern Russia can be additionally characterized as follows.

Advantages:

- reliance on entrepreneurship to assist improvement of living standards: focus on corporate taxes;
- Tax system transparency: focus on direct taxes;
- Aiming at tax liability and reducing the burden on tax authorities: a focus on taxes that are independently initiated and calculated by economic entities;

Disadvantages:

- internal orientation of the tax burden and insufficient use of globalization opportunities: prevailing of tax revenues from domestic economic activity;
- low efficiency of tax administration and high risk-related component of the tax system: significant scale of tax evasion (shadow economy);
- inadequate adoption of the tax burden by economic entities: investment disincentive and undermining performance motivation.

Thus, in accordance with available tax classifications, the tax burden in Russia is unsuitable, and despite sustainability of the tax system and the replenishment of government budget, the aggregate tax burden on economic entities is rather high and attitude is negative.

4 Conclusion

Summarizing, we can conclude that the hypothesis under suggestion is proofed. We substantiated that the tax burden in modern Russia is unreasonably high in comparison with the level that is acceptable for economic entities, despite lower tax rates against other countries (as illustrated by Germany). Unsuitability of the tax burden with a high probability may be a constraint in overcoming crisis impact on the Russian economy.

We have made a macroeconomic description of the tax burden in modern Russia, which allowed revealing both its advantages (sustainability, economic viability for the state, assistance in raising of living standards, transparency, commitment to tax liability and reducing the burden on tax authorities), and its disadvantages (internal orientation, low efficiency, high risk-related component, insufficient approval of tax burden by economic entities). Overcoming disadvantages will give prospects to upgrade the tax system in Russia.

References

Andrejovská, A., & Hudáková, M. (2016). Classification of EU countries in the context of corporate income tax. *Acta Universitatis Agriculturae et Silviculturae Mendelianae Brunensis, 64*(5), 1699–1708.

Bauman, M. P., & Shaw, K. W. (2016). Balance sheet classification and the valuation of deferred taxes. *Research in Accounting Regulation, 28*(2), 77–85.

Duan, T., Ding, R., Hou, W., & Zhang, J. Z. (2018). The burden of attention: CEO publicity and tax avoidance. *Journal of Business Research, 87*, 90–101.

Fang, H., Bao, Y., & Zhang, J. (2017). Asymmetric reform bonus: The impact of VAT pilot expansion on China's corporate total tax burden. *China Economic Review, 46*, S17–S34.

Federal Service of State Statistics of the Russian Federation (Rosstat). (2018). *Regions of Russia. Social and economic performances.* URL: http://www.gks.ru/wps/wcm/connect/rosstat_main/rosstat/en/statistics/publications/catalog/doc_1138623506156. Data accessed: 04/06/2018.

Garcia, M. M., & von Haldenwang, C. (2016). Do democracies tax more? Political regime type and taxation. *Journal of International Development, 28*(4), 485–506.

Gashenko, I. V., Zima, Y. S., Stroiteleva, V. A., & Shiryaeva, N. M. (2018). The mechanism of optimization of the tax administration system with the help of the new information and communication technologies. *Advances in Intelligent Systems and Computing, 622*, 291–297.

He, J. (2017). Effect of tax burden on income management. In *4th International Conference on Industrial Economics System and Industrial Security Engineering, IEIS 2017*, 8078567.

Heim, J. J. (2016). Do government stimulus programs have different effects in recessions, or by type of tax or spending program? *Empirical Economics, 51*(4), 1333–1368.

International Monetary Fund. (2018). *Shadow economies around the world: What did we learn over the last 20 years?* URL: https://www.imf.org/~/media/Files/.../WP/.../wp1817.ashx. Data accessed: 04.06.2018).

Kim, D. (2018). Projected impacts of federal tax policy proposals on mortality burden in the United States: A microsimulation analysis. *Preventive Medicine, 111*, 272–279.

Popkova, E. G., Bogoviz, A. V., Lobova, S. V., & Romanova, T. F. (2018a). The essence of the processes of economic growth of socio-economic systems. *Studies in Systems, Decision and Control, 135*, 123–130.

Popkova, E. G., Bogoviz, A. V., Ragulina, Y. V., & Alekseev, A. N. (2018b). Perspective model of activation of economic growth in modern Russia. *Studies in Systems, Decision and Control, 135*, 171–177.

The Russian Federation Tax Service. (2018). *Tax analytics: The revenue structure of the Russian consolidated budget*. URL: https://analytic.nalog.ru/portal/index.ru-RU.htm. Data accessed: 04/06/2018.

World Bank. (2018a). *Doing business 2018: Germany*. URL: http://www.doingbusiness.org/~/media/WBG/DoingBusiness/Documents/Profiles/State/DEU.pdf. Data accessed: 04.06.2018.

World Bank. (2018b). *Doing business 2018: Russian Federation*. URL: http://russian.doingbusiness.org/~/media/WBG/DoingBusiness/Documents/Profiles/State/RUS.pdf. Data accessed: 04.06.2018.

Part II
Theory and Practice of the Tax System Operation

Principles and Methods of Taxation

Irina V. Gashenko, Yuliya S. Zima and Armenak V. Davidyan

Abstract *Purpose* The purpose of the paper is to study available methods and principles of taxation and substantiate a necessary introduction of technical innovation principle. *Methodology* The methodology background is systematization and classification methods. Also, we applied ones of regression and correlation analysis. *Results* We revealed available methods of taxation and presented that the selection should be based on the tax principles. These principles should be proceeded from the peculiarities and priorities of social and economic system development and be divided into three categories: the constitutional principles of taxation, the principles of tax process management and the principles of resolving tax disputes. We managed to illustrate it owing to proprietary structural and logical scheme of principles and methods of taxation in the tax system. Collectively, they are aimed at the tax system improvement. Therewith, application of technical innovation plays an important role. *Recommendations* On the base of revealed close and direct dependence of the tax system competitiveness on the extent of E-services integration in the tax practice through the countries worldwide in 2018, it is advised to add the technical innovation principle to the principles of the tax process management.

Keywords Tax principles · Methods of taxation · Tax process
Tax system

JEL Classification E62 · H20 · K34

I. V. Gashenko (✉) · Y. S. Zima · A. V. Davidyan
Rostov State Economic University (Rostov Institute of National Economy),
Rostov-on-Don, Russian Federation
e-mail: gaforos@rambler.ru

Y. S. Zima
e-mail: zima.julia.sergeevna@gmail.com

A. V. Davidyan
e-mail: dav_121192@mail.ru

© Springer Nature Switzerland AG 2019
I. V. Gashenko et al. (eds.), *Optimization of the Taxation System: Preconditions, Tendencies, and Perspectives*, Studies in Systems, Decision and Control 182,
https://doi.org/10.1007/978-3-030-01514-5_4

1 Introduction

One of the most important features of any tax system is fundamental principles and applicable methods of taxation herein. Together, they are intended to improve the tax system. In this regard, successful performance of the tax mechanism in social and economic system depends on a reasonable selection and practical implementation of taxation principles and methods. In modern economic science and practice, the tax systems are roughly divided in developed and developing, since each of these categories have common patterns and similar features.

In this context, the countries with a transitional (emerging market) economy like modern Russia, attract the most interest, because they occupy an intermediate position between developed and developing countries and exhibit characteristic features of both categories. To reveal the peculiarities of the tax system management in transitional economy countries, which are understudied and simultaneously the most specific and scientifically attractive, in this paper we focus on the experience of the tax system management in modern Russia.

A preview of available scientific studies and regulatory documents showed that the whole complex of taxation principles does not include the requirement for application of technical innovation. Given that, in most areas of modern economic activity, including entrepreneurship and receiving other public services (except for taxation as a special service), we observe an increasing efficiency after integration of technical innovation that lead to improvement of services.

Therefore, in this study we put forward a hypothesis that in terms of the development of information-oriented society and the establishment of digital economy, the need for informatization (including digitalization) of the tax process is growing more urgent, since out-of-date technologies prevent full implementation of efficiency capacity and, consequently, provision of suitable process. The purpose of the paper is to study available methods and principles of taxation and substantiate a necessary introduction of technical innovation principle.

2 Materials and Method

Taxation methods are particularly investigated and discussed in the papers by experts such as Casajus (2015), Chen et al. (2018), Gashenko et al. (2018), Hemel (2018), Oishi et al. (2018), Vasilev (2015), Taxation principles are stated and analyzed in the publications by scientists such as Da Silva Santos Filho and Ferreira, PRA (2017), Keen and Mullins (2016), Konvisarova et al.(2016), Lee-Makiyama and Verschelde(2017), Popkova et al. (2018a, b).

The complex content analysis of many scientific studies and publications demonstrated that taxation principles become variously denominated and interpreted in the papers by different authors and tax systems of different countries worldwide. It causes heterogeneity and inconsistency of the available scientific

Table 1 Extent of E-services integration in the tax practice and the tax system competitiveness in different countries worldwide in 2018

Country	Extent of E-services integration in the tax practice, percent (x)	The tax system competitiveness, points from 0 to 100 (position) (y)
Estonia	81.20	100
New Zealand	66.56	88.7
Switzerland	54.35	85.2
Turkey	40.92	73.7
UK	52.54	70.8
Norway	47.27	70.7
Canada	53.50	69.1
Japan	54.27	66.8
Germany	44.76	66.6
Mexico	33.24	62.2
USA	48.00	55.1
Italy	29.91	47.7
France	28.61	43.4
Russia	20.11	34.4

Source Drawn up by authors on the data (Waseda-IAC 2018; Tax Foundation 2018)

knowledge of taxation principles and makes more urgent an issue of their systematization, which is the subject of this study.

Therefore, the framework methodology is systematization and classification methods. To test the hypothesis, the authors employ regression and correlation analysis methods that allow revealing the nature and strength of relation between the extent of E-services integration in the tax practice (x) (contained in Waseda-IAC data), as an index of technical innovation, and the tax system competitiveness (y) (calculated by the Tax Foundation) in the end of 2017-beginning of 2018 in different countries worldwide (Table 1).

3 Results

Traditionally, the following taxation methods are established. They are understood as a procedure to fix the tax rate depend on the tax base amount:

- method of equal taxation: equal (fixed) amount of tax, regardless of the tax base amount;
- method of proportional taxation: equal (fixed) tax rate, regardless of the tax base amount, while the tax amount increases with the tax base growth;
- regressive taxation method: decrease in the tax rate with the tax base growth;
- progressive taxation method: increase in the tax rate with the tax base growth.

Taxation methods is selected individually for each type of tax; it depends on the tax principles, which composition, expression and interpretation are specified due to peculiarities and priorities of social and economic system development. For example, negative attitude of taxpayers to government initiatives to replace the proportional taxation method with the progressive one, when levying the personal income tax and the profit tax in modern Russia, is the reason to maintain a current (proportional) taxation method (VCIOM 2018).

As follows from classification, we have identified the main categories of taxation principles that are stated in most scientific papers and peculiar to the most tax systems worldwide, which in turn contain varieties of taxation principles studied in this paper in the context of tax practice in modern Russia.

The first category is constitutional principles enunciated in the Constitution. They establish fundamentals of taxation and constitute rights, freedoms and obligations of the tax system participants and include:

- the principle of tax legitimacy: the tax system operation should be in accordance with the law, and taxpayers should pay only taxes and fees under terms that are enshrined in law;
- the principle of universal coverage and equality of taxation conditions: all people should fulfill the requirements of the tax legislation in full on general, equal conditions (that is, on conditions of equality before the law).

The second category is the principles of the tax process management enshrined in the Tax Code. They define the tax process parameters and the tax practice priorities in social and economic system and include:

the principle of fair taxation: taxpayers should have a real opportunity to pay taxes (commensurate with their position). We should achieve a social justice (fair redistribution of revenues in the society and differentiation of tax conditions for various categories of taxpayers);

the principle of economic relevance of taxation: the establishment of taxes should not be arbitrary, but should be conditioned by objective need (the target nature of taxation);

the principle of the tax system unity: rights, freedoms and obligations to pay taxes are applicable to all territories within the country (upon admission to pursue own tax policy of these territories in accordance with the nation-wide tax policy);

the principle of the tax system transparency: rights, freedoms and obligations in the field of taxation should be intelligible to all taxpayers, therefore taxation conditions should be simple, precise and sustainable.

The third category is the principles of resolving tax disputes. They are enshrined in the Tax Code and in the Constitutional Court rulings and intended to set priorities in resolving disputes arising in the tax process. These are:

- principle of access to courts to defend the tax interests: if the tax rights are violated, everyone has the right to apply for their protection;
- principle of priority of taxpayers' interests: in resolving tax disputes (inconsistency of tax legislation and tax practice) judgements should be made in favor of taxpayers.

In order to evaluate the practicability of supplementing above-noted principles with potentially high-demanded principle of technical innovation in the tax practice, let us turn to the results of our regression and correlation analysis (Fig. 1).

As can be seen from Fig. 1, increase in the extent of E-services integration in the tax practice of countries worldwide in 2018 of 1% contributes to the growth in the competitiveness of their tax systems of 1.0175 points with a correlation of 83.26%. This evidences close and direct dependence of the tax system competitiveness on the extent of E-services integration in the tax practice.

Therefore, observance of technical innovation principle is a central premise and an essential condition to ensure suitability of tax systems in modern world. Therefore, we advise this principle to be included in the category of the tax process management. Pursuant thereto, we constructed the following structural and logical scheme of principles and methods of taxation (Fig. 2).

As can be seen from Fig. 2, the principles of taxation should be interrelated to improve the tax system. Their core is the constitutional principles of taxation, which in turn are basic for the principles of the tax process management. The latter constitute a background for the principles of resolving tax disputes. The principles of taxation are established in accordance with the peculiarities and priorities of social and economic system development and define the methods of taxation.

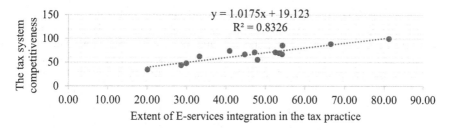

Fig. 1 Regression curve displaying the dependence of the tax system competitiveness on the extent of E-services integration in the tax practice through the countries worldwide in 2018. *Source* Calculated and compiled by the authors

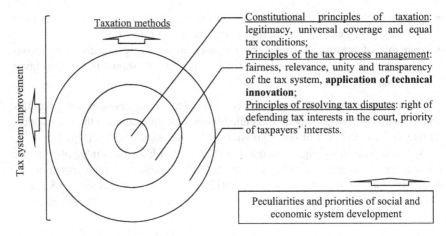

Fig. 2 Structural and logical scheme of taxation principles and methods in the tax system. *Source* Compiled by the authors

4 Conclusion

Thus, we revealed the methods of equal, proportional, regressive and progressive taxation. The method should be selected due to the principles of taxation. These principles are based on the peculiarities and priorities of social and economic system development and are divided into three categories: the constitutional principles of taxation, the principles of the tax process management and the principles of resolving tax disputes. We managed to illustrate it owing to proprietary structural and logical scheme of principles and methods of taxation in the tax system.

Collectively, they are aimed at the tax system improvement. Therewith, application of technical innovation plays an important role. The close and direct dependence of the tax system competitiveness on the extent of E-services integration in the tax practice through the countries worldwide in 2018 proofed the hypothesis and became the basis to make a proprietary suggestion on introduction of technical innovation principle into the principles of the tax process management.

References

Casajus, A. (2015). Monotonic redistribution of non-negative allocations: A case for proportional taxation revisited. *Economics Letters, 136,* 95–98.

Chen, B.-L., Hsu, M., & Hsu, Y.-S. (2018). Progressive taxation and macroeconomic stability in two-sector models with social constant returns. *Journal of Economics/ Zeitschrift fur Nationalokonomie, 2*(1), 1–18.

Da Silva Santos Filho, I., & Ferreira, P. R. A. (2017). Fundamental principles of environmental taxation. *Veredas do Direito, 14*(29), 125–151.

Gashenko, I. V., Zima, Y. S., Stroiteleva, V. A., & Shiryaeva, N. M. (2018). The mechanism of optimization of the tax administration system with the help of the new information and communication technologies. *Advances in Intelligent Systems and Computing, 622,* 291–297.

Hemel, D. J. (2018). Federalism as a safeguard of progressive taxation. *New York University Law Review, 93*(1), 1–57.

Keen, M.,& Mullins, P. (2016). International corporate taxation and the extractive industries: Principles, practice, problems. In *International Taxation and the Extractive Industries: Resources Without Borders* (pp. 11–4). UK: Taylor and Francis.

Konvisarova, E. V., Stihiljas, I. V., Koren, A. V., Kuzmicheva, I. A., & Danilovskih, T.E. (2016). Principles of profit taxation of commercial banks in Russia and Abroad. *International Journal of Economics and Financial Issues, 6*(8), 189–192.

Lee-Makiyama, H., & Verschelde, B. (2017). OECD BEPS: Reconciling global trade, taxation principles and the digital economy. In *The Challenge of the Digital Economy: Markets, Taxation and Appropriate Economic Models* (pp. 55-68). Berlin: Springer International Publishing.

Oishi, S., Kushlev, K., & Schimmack, U. (2018). Progressive taxation, income inequality, and happiness. *American Psychologist, 73*(2), 157–168.

Popkova, E. G., Bogoviz, A. V., Lobova, S. V., & Romanova, T. F. (2018a). The essence of the processes of economic growth of socio-economic systems. *Studies in Systems, Decision and Control, 135,* 123–130.

Popkova, E. G., Bogoviz, A. V., Ragulina, Y. V., & Alekseev, A. N. (2018b). Perspective model of activation of economic growth in modern Russia. *Studies in Systems, Decision and Control, 135,* 171–177.

Russian Public Opinion Research Center (VCIOM). (2018). Constraining Factors to Business and Opportunities of Economic Growth–2018. https://wciom.ru/index.php?id=236&uid=9108. Data accessed: 05.06.2018.

Tax Foundation. (2018). 2017 International Tax Competitiveness Index. https://taxfoundation.org/2017-international-tax-competitiveness-index/. Data accessed 02.06.2018.

Vasilev, A. (2015). Welfare gains from the adoption of proportional taxation in a general-equilibrium model with a grey economy: The case of Bulgaria's 2008 flat tax reform. *Economic Change and Restructuring, 48*(2), 169–185.

Waseda-IAC. (2018). International e-Government Rankings–2017. https://www.waseda.jp/top/en-news/53182. Data accessed: 05.06.2018.

Tax Administration and Control. Tax Policy of the State

Irina V. Gashenko, Yuliya S. Zima and Armenak V. Davidyan

Abstract *Purpose* The purpose of this paper is to study public administration of the tax system, in particular, development and implementation of public tax policy and execution of tax administration and control, as well as to evaluate the suitability of the tax system management in modern Russia. *Methodology* The author elaborated) and employed the method of qualitative evaluation of sustainable management of the tax system at different stages in the context of economic cycle phases. *Results* We have drawn up a conceptual model for public management of the tax system that is typical for most modern countries, including Russia. It demonstrated that public management is rather complicated and is implemented gradually and cyclically, including macro- and geo-economic monitoring and defining general purposes of public economic policy, developing and implementing tax policy, and executing tax administration and control. A qualitative evaluation of suitable management of the Russian tax system in 2006–2017 with the help of proprietary method had been carried out. *Recommendations* We revealed that suitable public management of the tax system is not achieved in modern Russia. To tackle with this issue and improve the tax management, it is advised to upgrade the practical model of this process through supplementing taxpayer-state feedback.

Keywords Tax system management · Tax administration and control
Public tax policy · Russia

JEL Classification E62 · H20 · K34

I. V. Gashenko (✉) · Y. S. Zima · A. V. Davidyan
Rostov State Economic University (Rostov Institute of National Economy),
Rostov-on-Don, Russian Federation
e-mail: gaforos@rambler.ru

Y. S. Zima
e-mail: zima.julia.sergeevna@gmail.com

A. V. Davidyan
e-mail: dav_121192@mail.ru

© Springer Nature Switzerland AG 2019   41
I. V. Gashenko et al. (eds.), *Optimization of the Taxation System: Preconditions,
Tendencies, and Perspectives*, Studies in Systems, Decision and Control 182,
https://doi.org/10.1007/978-3-030-01514-5_5

1 Introduction

In modern world the public tax system continually undergoes alterations due to impact of two forces. The first force is an exposure of market economy to cyclical fluctuations. Dynamic macroeconomic situation constantly transforms the tax process thereby changing taxation opportunities (group and structure of taxpayers, tax base amount, etc.). The second force is the globalization impact on modern social and economic systems. The external environment of these systems lays down their variable tax demands related to maintaining their sustainability in terms of crisis or building tax capacity in stable conditions or creation of reserves within economic growth.

In this regard, suitable tax system is featured by its static state at a certain point. To maintain it in the long term (over a long period of time), it is necessary to sustain its high flexibility and adaptivity to affecting forces. The methods to achieve these features of the tax system are public administration tools—development and implementation of tax policy, tax administration and control. The study of these tools becomes especially urgent within pursuance of the tax system suitability.

The working hypothesis of this study rests on the assumption that suitable tax public management is not achieved in modern Russia. The purpose of this paper is to study public administration of the tax system, in particular, development and implementation of public tax policy and execution of tax administration and control, as well as to evaluate the suitability of the tax system management in modern Russia.

2 Materials and Method

The review of the literature shown that the issues of tax public management have been particularly studied by modern scientists. The topic of tax administration and control is investigated in the papers by Gashenko et al. (2018), Keen and Slemrod (2017), Olivares (2018), Rubio et al. (2017), Solehzoda (2017), Tjen and Evans (2017. The peculiarities of theory and practice in development and implementation of public tax policy are discussed in the publications by Baiardi et al. (2018), Colombo and Caldeira (2018), Fuest (2018), Popkova et al. (2018a, b), Vasilev (2018).

Therewith, despite the abundance of fundamental and empirical studies that stress the importance of upgrading tax system management, the methodological issues to achieve and evaluate are insufficiently investigated. Therefore, for the purposes of this research, we developed a proprietary method for qualitative evaluation of suitable tax system management at different stages in the context of economic cycle phases, which is described in Table 1.

As can be seen from Table 1, the method in question stipulates a qualitative evaluation, when the experts assign qualitative values to the evaluation criteria such

Table 1 The method of qualitative evaluation of suitable tax system management at different stages in the context of economic cycle phases

Stage of the tax system management	Evaluation criteria	Performances to make a criterion evaluation		
		In terms of crisis and stagnation	In terms of growth	In terms of stability
Preparation: tax policy development	Degree of correlation between the tax policy in current macroeconomic situation and common guidelines of public economic policy	Policy of reducing expenditures (budget gap)	Maximum revenues policy (budget surplus)	Policy to adjust revenues and expenditures (budget balance)
Process: tax administration and control	Efficiency of tax administration and control	– Value of tax revenues; – Shadow economy scale; – Efficiency (ratio of revenues to shadow economy)		
Result: outcomes of the tax system management	Extent of achieving initial goals of tax system management (tax policy)	Level and rate of economic growth (economic recovery)	Budget surplus value (accumulated reserves)	Living standards of the population

Source Drawn by authors

as "low", "average" and "high". The tax system management may be deemed suitable only with "high" values for all evaluation criteria.

3 Results

On the results of studying public management of the tax system, we developed its conceptual model (Fig. 1).

As can be seen from Fig. 1, the subject of the tax system management is the state (represented by relevant tax regulators, such as the Federal Tax Service of the Russian Federation, etc.), and the object of taxation is the tax process. The latter is understood as the one to carry out economic activities that are liable to taxation (the participants are taxpayers).

Public management of the tax system starts with (1 stage) the monitoring of macro- and geo-economic situation and the definition of general guidelines of public economic policy, where the tax system management is a part. Thereunder (stage 2), the tax policy is developed and implemented as one of two management tools. The following logic is applied to select the type of tax policy:

- In terms of economic crisis and stagnation, it is expedient to implement the tax policy of reducing expenditures (budget deficit), the purpose thereof should be achieving an economic growth (economic recovery);

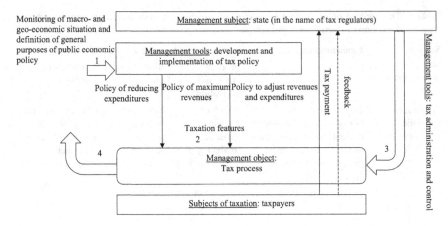

Fig. 1 The model of the tax system public management. *Source* Developed by authors

- In terms of growth, it is expedient to implement the maximum revenue tax policy (budget surplus), the purpose thereof should be achieving a budget surplus (accumulation of reserves in the event of crisis);
- In terms of stability, it is expedient to implement the tax policy to adjust revenues and expenditures (budget balance), the purpose thereof should be achieving an improvement in the living standards of the population (GDP per capita).

With the help of the tax policy it's possible to establish taxation features such as the ratio of direct and indirect taxes, the ratio of taxes at different levels of the fiscal system, the distribution of the tax burden between taxpayers, the regulation of tax benefits, tax composition, taxation objects, tax base, tax rates, etc.

Then follows (Stage 3) the tax administration and control aimed at maximizing the amount of tax revenues and minimizing the shadow economy scale (tax evasion)—the ratio of these performances decides its effectiveness. The payment of taxes is affected by these management tools.

Also, scientifically and theoretically a model of public administration of the tax system should provide a feedback through the taxpayers can ask questions to the state and raise objections and proposals on improving the tax system management. In Fig. 1 we depicted feedback in a dashed arrowhead line, because practically it may be absent, which is typical for modern Russia. In the model there is a fourth stage, where we return to the first stage and terminate the cycle of the tax system public management.

The results of qualitative evaluation of suitable tax system management at different stages in the context of economic cycle phases in Russia in 2006–2017 on the Rosstat data with the help of proprietary method are illustrated in Fig. 2. To reveal economic cycle phases, we joined two diagrams—a diagram displaying the dynamic pattern of Russian GDP, and a diagram displaying the dynamic pattern of the consolidated budget balance.

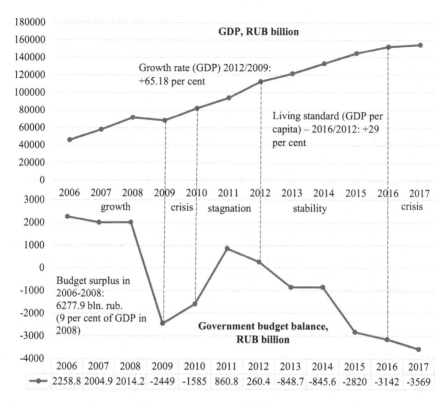

Fig. 2 The results of qualitative evaluation of sustainable tax system management in Russia in 2006–2017. *Source* Developed by authors on the data (Rosstat 2018)

As can be seen from Fig. 2, the term of 2006–2008 was a phase of growth (increase in GDP). This period the country implemented the tax policy of the maximum revenues that lead to the surplus of the consolidated budget in the amount of 6277.9 billion rubles (9% of GDP in 2008). In 2009, there was a crisis (a sharp decline in GDP), which in 2010–2012 was followed by stagnation of the Russian economy. Over this period, the country had been pursuing the tax policy of reducing expenditures (leading to a budget gap), which resulted in GDP growth of 65.18% in 2012 against 2009.

In 2006–2008 the economy was stable (a slight increase in GDP). At that moment, the tax policy to adjust revenues and expenditures took place, which resulted in a 29% increase in the living standard (GDP per capita) in 2016 against 2012. Consequently, over 2006–2017 the extent of correlation between the tax policy and current macroeconomic situation and general guidelines of public economic policy and the extent to achieve initial purposes of the tax system (tax policy) management were very high.

However, in 2017 tax revenues amounted to 32.80% of GDP (World Bank 2018), while the share of the shadow economy in Russia was 38.48% of GDP

(International Monetary Fund 2018). It highlights a very low efficiency of tax administration and control (32.80/38.48 = 0.85, that is less than 1), which is the reason for unsuitability of tax management in modern Russia, despite its high flexibility and adaptivity to changing macroeconomic situation and general guidelines of public economic policy.

4 Conclusion

We can conclude that the process of tax system public management is rather complicated and is carried out in four subsequent stages generating a closed cycle. The first (preparatory) stage is the monitoring of macro- and geo-economic situation and the definition of general guidelines of public economic policy. The second stage is connected with the development and implementation of public tax policy (a tool for tax management), the third—with the implementation of tax administration and control (a tool for tax management). The fourth stage involves a return to the first stage and ensure the cyclical nature of this process.

We uncovered that the flaw in the developed conceptual model of public management of the tax system specific to most modern countries of the world, including Russia, is the absence of both direct and indirect (feedback) interaction between subjects of the tax process—the state and taxpayers. This can cause separateness of the tax system management from current trends in economic practices of the country and leads to tax evasion, i.e. to constituting the shadow economy, which is observed in modern Russia.

We proofed the working hypothesis and revealed that suitable public management of the tax system is not achieved in modern Russia. Despite the fact that at the stage of developing the tax policy and summarizing the results of the tax system management, high efficiency (correlation between economic cycle phases) is achieved, at an intermediate stage of tax administration and control—there is very low efficiency caused by the excess of shadow economy over tax revenues in the consolidated budget of the Russian Federation. To solve this issue and upgrade the tax system management in modern Russia, it is advised to improve the practical model of this process as a whole (and not only at the stage of tax administration and control), through supplementing taxpayer-state feedback.

References

Baiardi, D., Profeta, P., Puglisi, R., & Scabrosetti, S. (2018). Tax policy and economic growth: Does it really matter? *International Tax and Public Finance, 2*(10), 1–35.

Colombo, J. A., & Caldeira, J. F. (2018). The role of taxes and the interdependence among corporate financial policies: Evidence from a natural experiment. *Journal of Corporate Finance, 50*, 402–423.

Federal Service of State Statistics of the Russian Federation (Rosstat). (2018). Russia in Figures. http://www.gks.ru/wps/wcm/connect/rosstat_main/rosstat/ru/statistics/publications/catalog/doc_1135075100641. Data accessed: 04.06.2018.

Fuest, C. (2018). Tax policy as an instrument for protectionism? | [Steuerpolitik als Mittel des Protektionismus?]. *Wirtschaftsdienst, 98*(2269), 4–7.

Gashenko, I. V., Zima, Y. S., Stroiteleva, V. A., & Shiryaeva, N. M. (2018). The mechanism of optimization of the tax administration system with the help of the new information and communication technologies. *Advances in Intelligent Systems and Computing, 622,* 291–297.

International Monetary Fund. (2018). Shadow Economies Around the World: What Did We Learn Over the Last 20 Years? https://www.imf.org/~/media/Files/.../WP/.../wp1817.ashx. Data accessed: 04.06.2018.

Keen, M., & Slemrod, J. (2017). Optimal tax administration. *Journal of Public Economics, 152,* 133–142.

Olivares, B. D. O. (2018). Technological innovation within the Spanish tax administration and data subjects' right to access: An opportunity knocks. *Computer Law and Security Review, 34* (3), 628–639.

Popkova, E. G., Bogoviz, A. V., Lobova, S. V., & Romanova, T. F. (2018a). The essence of the processes of economic growth of socio-economic systems. *Studies in Systems, Decision and Control, 135,* 123–130.

Popkova, E. G., Bogoviz, A. V., Ragulina, Y. V., & Alekseev, A. N. (2018b). Perspective model of activation of economic growth in modern Russia. *Studies in Systems, Decision and Control, 135,* 171–177.

Rubio, E. V., González, P. E. B., & Alaminos, J. D. (2017). Relative efficiency within a tax administration: The effects of result improvement. *Revista Finanzas y Política Económica, 9* (1), 135–149.

Solehzoda, A. (2017). Information technologies in the tax administration system of VAT. *Journal of Advanced Research in Law and Economics, 8*(4), 1340–1344.

Tjen, C., & Evans, C. (2017). Causes and consequences of corruption in tax administration: An Indonesian case study. *Journal of Tax Research, 15*(2), 243–261.

Vasilev, A. (2018). Optimal fiscal policy with utility-enhancing government spending, consumption taxation and a common income tax rate: The case of Bulgaria. *Review of Economics, 69* (1), 43–58.

World Bank. (2018). Doing Business 2018: Russian Federation. http://russian.doingbusiness.org/~/media/WBG/DoingBusiness/Documents/Profiles/Country/RUS.pdf. Data accessed: 04.06.2018.

Standard and Special Tax Treatments. A Comparative Analysis

Irina V. Gashenko, Yuliya S. Zima and Armenak V. Davidyan

Abstract *Purpose* The purpose of this study is to carry out a comparative analysis of standard and special tax treatments in modern Russia and define the attractiveness of special tax treatments for economic entities. *Methodology* In this paper we apply a specially developed method of criterion evaluation of the attractiveness of special tax treatments for economic entities. *Results* The authors conclude that in modern Russia there are four special tax treatments—a simplified tax system, a patent tax system, a unified agricultural tax and a unified tax on imputed income. They give tax preferences to small and medium business entities such as a reduced tax rate of 5% on business income (in comparison with a 20% profit tax) and a slight simplification of tax reporting. Their advantage is a quite wide (with the prospect of expansion) coverage of priority entrepreneurial activity, and disadvantages are the necessity to pay many other taxes, the continuing high complexity of tax reporting, despite some simplification, as well as difficult transition and necessary regular confirmation of the right to use a special tax treatment. *Recommendations* Insignificant advantages and multiple disadvantages cause low attractiveness of special tax treatments for economic entities. Therefore, in order to upgrade the tax system of modern Russia, it is advised to increase the value of tax preferences and raise the simplicity, accessibility and convenience of special tax treatments.

Keywords Standard tax treatment · Special tax treatments · Attractiveness of tax treatment · Russia

JEL Classification E62 · H20 · K34

I. V. Gashenko (✉) · Y. S. Zima · A. V. Davidyan
Rostov State Economic University (Rostov Institute of National Economy),
Rostov-on-Don, Russian Federation
e-mail: gaforos@rambler.ru

Y. S. Zima
e-mail: zima.julia.sergeevna@gmail.com

A. V. Davidyan
e-mail: dav_121192@mail.ru

© Springer Nature Switzerland AG 2019   49
I. V. Gashenko et al. (eds.), *Optimization of the Taxation System: Preconditions,
Tendencies, and Perspectives*, Studies in Systems, Decision and Control 182,
https://doi.org/10.1007/978-3-030-01514-5_6

1   Introduction

Taxation is one of the key tools of public promotion of not only quantitative (GDP growth rate), but also qualitative (GDP structure) economic growth. For this purpose, the state establishes special tax treatments that provide tax preferences for priority economic activities in addition to standard ones. It is a difficult task. Firstly, it is necessary to provide these preferences within special tax treatments that are beneficial for economic entities and not too much burdensome for the state.

Secondly, it is required to establish transition barriers to special tax treatments that make their available to target economic entities and not admit non-target economic entities. Thirdly, it is important to ensure performances of special tax treatments that they are to be universal and suitable for the vast majority of target economic entities, since the establishment of various special tax treatments will only complicate the tax system, violating the principle of transparency.

Operation of special tax treatments is justified only in the event of their high attractiveness for economic entities, which is one of the essential conditions to achieve suitable tax system. Due to revealed high level of the shadow economy that evidences mass tax evasion, we put forward a hypothesis that special tax treatments are low-attractive for economic entities in modern Russia.

The purpose of this study is to carry out a comparative analysis of standard and special tax treatments in modern Russia and define the attractiveness of special tax treatments for economic entities.

2   Materials and Method

Both from a conceptual point of view and an empirical one, as illustrated by different countries of the world, tax treatments are investigated in numerous papers by modern authors, among them the following publications: Bumane and Vodolagins (2017), Choudhury (2018), Fréret and Maguain (2017), Gordon et al. (2016), Jacob (2018), Korotin et al. (2017), Lee and Mason (2015), Leonardi, (2017), Popkova et al. (2018a, b), Rietzler et al. (2017), Rosenblatt and Cabral (2017), Ruiz (2017), Stiegler et al. (2016) and Tufetulov et al. (2015).

However, the methodological issues to evaluate the attractiveness of special tax treatments for economic entities are not adequately covered in above-mentioned scientific papers. To fill this gap, in this paper we apply a specially developed method to evaluate the attractiveness of special tax treatments for economic entities by following criteria:

- value: the amount of tax benefits;
- availability: simple transition to a special treatment;
- completeness: special tax treatment coverage of all priority economic activities;
- convenience: elimination of all other taxes and fixing only one tax that is provided by a special tax treatment;

- Simplicity: difficult tax reporting.

This method assigns values in points from 1 to 10 to the listed criteria, where 1 is the least attractive, and 10 is the most attractive. Evaluation results are advised to be presented by a pentagon-shaped diagram, where each angle reflects the value assigned to the relevant criterion. Through this, it is easy to define advantages and disadvantages of special tax treatments.

3   Results

The results of studying taxation peculiarities of business entities in Russia as of 2018 showed that they should pay the following taxes and fees in accordance with standard tax treatment depend on the specific nature of their economic activity:

- tax on personal income (obligatory);
- insurance contributions (obligatory);
- profit tax (obligatory);
- value added tax (obligatory, but with possible tax rate of 0% for certain categories of goods);
- property tax (if available any property);
- land tax (if available any land in property);
- transport tax (if available any vehicles in property).

In Russia there are two categories of business entities: legal entities (companies, partnerships, institutions, etc.) and natural entities (individual enterprises). Individual enterprises fall into category of small and medium enterprises, for which various special tax treatments are provided. The results of our comparative analysis of available tax treatments in modern Russia are presented in Table 1.

Table 1 Comparative analysis of standard and special tax treatments in Russia in 2018

Tax treatment	Business income tax	Tax rate	Other taxes
Standard tax treatment	Profit tax (difference between revenues and expenditures)	20%	Yes
Simplified tax treatment	Income tax	6%	Yes
	Or tax on difference between revenues and expenditures	15%	Yes
Patent tax system	Patent	Payment of patent value	Only insurance contributions, land tax and transport tax
Unified agricultural tax	Difference between revenues and expenditures	6%	Yes
Unified tax on imputed income	Unified tax on imputed income	15%	Yes

Source Drawn up by authors on the data (Federal Tax Service of the Russian Federation 2018)

As can be seen from Table 1, standard tax treatment assumes the payment of business tax in the form of income tax at 20% rate on the difference of revenues and expenditures with obligatory payment of all taxes and fees listed above. Additionally, there are four special tax treatments.

The first is a simplified tax system. The conditions for transition to this special tax treatment is observance of one of three criteria: an income of no more than 150 million rubles per year, staff number of no more than 100 people, depreciated book value of fixed assets of no more than 150 million rubles. In this case, all taxes provided by standard tax treatment are paid, except for the profit tax, which is replaced by income tax (6%) or tax on the difference of revenues and expenditures (15%) at entrepreneur's option.

The second is the patent tax system. The conditions for transition to this special tax treatment are the inclusion in strictly fixed activities (approved individually by each region) and the acquisition of a patent, which value depends on the type of activity and region of entrepreneurial activity pursuit. In this situation, only the patent value is paid (as an alternative to the profit tax), insurance contributions (obligatory), land tax (if available any property) and transport tax (if available any vehicles in property).

The third is a unified agricultural tax. The conditions for transition to this special tax treatment is the enterprise specialization in the production of agricultural goods, where the share of sale profit should make at least 70% of the total revenue structure of the enterprise. In this situation, all taxes provided by standard tax treatment are paid, except for the profit tax, which is replaced by a unified agricultural tax (difference of revenues and expenditures, 6%).

The fourth is a unified tax on imputed income. The conditions for transition to this special tax treatment are the inclusion in strictly fixed activities. In this situation, all taxes provided by standard tax treatment are paid, except for the profit tax, which is replaced by an imputed income tax (15%)—production of basic return by government-set activity, a deflator coefficient (allows taking into account an inflation impact) and a coefficient considering the peculiarities of entrepreneurship (downward).

In accordance with the above stated, we evaluated special tax treatments at 6 points, since most common (widely accessible) of them are a simplified tax system and a unified tax on imputed income—contemplate a slight decrease in the tax rate (by 5%) compared to standard tax treatment.

Preferential tax conditions for participants of special economic zones are also provided, and the issue of introducing special tax treatments for technopark participants is examined. Therefrom, it can be argued that the main categories of economic entities carrying out priority activity for the economy (small and medium enterprises, manufacturers of agricultural goods, guidelines of regional economy growth, sources of innovative economic development) are embraced of special tax treatments in modern Russia. Therefore, we assign 7 points to a criterion of "completeness".

The accessibility of special tax treatments in Russia is evaluated by us at 5 points, because it requires not only initial, but also a subsequent periodic

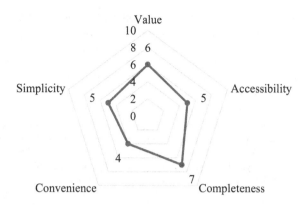

Fig. 1 Results of the criterial evaluation of the attractiveness of special tax treatments for economic entities in Russia in 2018 *Source* Compiled by the authors

documentary confirmation of legal use of special tax treatments by economic entities. The convenience gets 4 points, as almost all tax treatments assume maintaining other taxes, replacing only the profit tax. We evaluated the simplicity at 5 points, since tax reporting is submitted to various public institutions in periodically changing forms with limited application of e-reporting.

Thus, we assigned values to all criteria of attractiveness of special tax treatments for economic entities in Russia in 2018, which allowed us of representing evaluations by a pentagon-shaped diagram (Fig. 1).

As can be seen from Fig. 1, the attractiveness of special tax treatments for economic entities in Russia in 2018 can be described as low—the arithmetic mean of the corresponding criteria is 5.4 points of 10 possible. The reasons are inconvenience, complexity and inaccessibility of special tax treatments.

4 Conclusion

Following the results of the study, we concluded that in modern Russia there are four special tax treatments—a simplified tax system, a patent tax system, a unified agricultural tax and a unified tax on imputed income. They give tax preferences to small and medium business entities such as a reduced tax rate of 5% on business income (in comparison with a 20% income tax) and a slight simplification of tax reporting.

Their advantage is a quite wide (with the prospect of expansion) coverage of priority entrepreneurial activity, and disadvantages are the necessity to pay many other taxes, the continuing high complexity of tax reporting, despite some simplification, as well as difficult transition and necessary regular confirmation of the right to use a special tax treatment.

Insignificant advantages and multiple disadvantages cause low attractiveness of special tax treatments for economic entities. Therefore, in order to upgrade the tax system in modern Russia, it is advised to increase the value of tax preferences and raise the simplicity, accessibility and convenience of special tax treatments.

References

Bumane, I., & Vodolagins, D. (2017). Favourable tax regimes that constitute selective state aid from the perspective of the cjeu recent case-law. *European Research Studies Journal, 20*(3), 231–245.

Choudhury, M. A. (2018). The nature of well-being objective function in tax-free regime of ethico-economics. *Journal of Islamic Accounting and Business Research, 9*(2), 171–182.

Federal Tax Service of the Russian Federation. (2018). Tax Treatment Selection. https://www.nalog.ru/create_business/ip/in_progress/taxation_type_choice/. Data accessed: 06.06.2018.

Fréret, S., & Maguain, D. (2017). The effects of agglomeration on tax competition: evidence from a two-regime spatial panel model on French data. *International Tax and Public Finance, 24*(6), 1100–1140.

Gordon, R. N., Joulfaian, D., & Poterba, J. M. (2016). Choosing between an estate tax and a basis carryover regime: Evidence from 2010. *National Tax Journal, 69*(4), 981–1002.

Jacob, M. (2018). Tax regimes and capital gains realizations. *European Accounting Review, 27*(1), 1–21.

Korotin, V., Popov, V., Tolokonsky, A., Islamov, R., & Ulchenkov, A. (2017). *A Multi-criteria Approach to Selecting an Optimal Portfolio of Refinery Upgrade Projects Under Margin and Tax Regime Uncertainty* (Vol. 72, pp. 50–58). United Kingdom: Omega.

Lee, S.-H., & Mason, A. (2015). Are current tax and spending regimes sustainable in developing Asia? In *Inequality, Inclusive Growth, and Fiscal Policy in Asia* (pp. 202–234). UK: Taylor and Francis Inc.

Leonardi, R. (2017). The digital economy and the tax regime in the UK. In *The Challenge of the Digital Economy: Markets, Taxation and Appropriate Economic Models* (pp. 97–109). Berlin: Springer International Publishing.

Popkova, E. G., Bogoviz, A. V., Lobova, S. V., & Romanova, T. F. (2018a). The essence of the processes of economic growth of socio-economic systems. *Studies in Systems, Decision and Control, 135*, 123–130.

Popkova, E. G., Bogoviz, A. V., Ragulina, Y. V., & Alekseev, A. N. (2018b). Perspective model of activation of economic growth in modern Russia. *Studies in Systems, Decision and Control, 135*, 171–177.

Rietzler, K., Fuest, C., Kauder, B., … Schmidt, C. M., Schratzenstaller, M. (2017). How should the german tax regime be reformed? | [Wie sollte das Steuersystem in Deutschland reformiert werden?]. *Wirtschaftsdienst, 97*(6), 383–403.

Rosenblatt, P., & Cabral, R. T. P. (2017). Tax transparency regime in taxation of profits abroad (CFC Rules): Gaps and conflicts in Brazilian law | [Regime de transparência fiscal na tributação dos lucros auferidos no exterior (CFC Rules): Lacunas e conflitos no direito brasileiro]. *Brazilian Journal of International Law, 14*(2), 450–463.

Ruiz, M. A. R. (2017). Considerations on the establishment of a common tax regime for social economic entities | [Consideraciones en torno al establecimiento de un régimen tributario común para las entidades de la economía social]. *REVESCO Revista de Estudios Cooperativos, 125*, 187–212.

Stiegler, T., Wiesener, A. U., Bussian, A., & Schmid, J. (2016). The new German Tax regime for investment funds: An exemplar for Europe? In *Proceedings of the 28th International Business Information Management Association Conference—Vision 2020: Innovation Management, Development Sustainability, and Competitive Economic Growth* (pp. 2005–2013).

Tufetulov, A. M., Davletshin, T. G., & Salmina, S. V. (2015). Analysis of the impact of special tax regimes for small business financial results. *Mediterranean Journal of Social Sciences, 6*(1S3), 503–506.

Tax Burden and Mitigation of Tax Payments

Irina V. Gashenko, Yuliya S. Zima and Armenak V. Davidyan

Abstract *Purpose* The purpose of the paper is to study the essence of tax burden and mitigate tax payments and their peculiarities in modern Russia. *Methodology* The methodological unit of this paper is built on a systematic approach and is based on the application of methods of structural and comparative analysis, synthesis, induction, deduction and formalization, as well as a complex of methods to estimate the value (in absolute terms) and the extent (in relative terms) of tax burden. *Results* In course of the study it was revealed that tax burden in modern Russia is a relatively high, which likely causes a shadow economy. The aggregated tax burden on the Russian population as of the end of 2017-beginning of 2018 amounts to 32%. It is mainly due to the personal income tax with limited opportunities to be minimized. The aggregated tax burden on the Russian business as of the end of 2017-beginning of 2018 is 5.39% of sales revenue and 69.44% of before-tax profit. It is mainly due to the profit tax and the mineral extraction tax, with big opportunities to minimize payments including both specialization in non-resource branches of economy and selection of special tax treatment. *Recommendations* Taking into account a revealed imbalance in tax burden distribution between the business and the population towards an increased burden on the latter, it is advised to provide in modern Russia additional opportunities to reduce tax burden on the population through simplifying procedure to obtain benefits on the profit tax payment.

Keywords Tax burden · Tax opportunism · Minimization of tax payments
Modern russia

JEL Classification E62 · H20 · K34

I. V. Gashenko (✉) · Y. S. Zima · A. V. Davidyan
Rostov State Economic University (Rostov Institute of National Economy),
Rostov-on-Don, Russian Federation
e-mail: gaforos@rambler.ru

Y. S. Zima
e-mail: zima.julia.sergeevna@gmail.com

A. V. Davidyan
e-mail: dav_121192@mail.ru

© Springer Nature Switzerland AG 2019
I. V. Gashenko et al. (eds.), *Optimization of the Taxation System: Preconditions, Tendencies, and Perspectives*, Studies in Systems, Decision and Control 182,
https://doi.org/10.1007/978-3-030-01514-5_7

1 Introduction

Taxation is a flexible tool of public regulation of economy and an important force of economic systems operation. Within stability and economic growth, the interests to Marshall tax resources for successful public performance are prevailing, but in terms of crisis taxes are designed to be a mechanism for promoting investment and innovation activity in the economic system to seize decline, prevent longstanding stagnation and overcome crisis impact.

Macroeconomic development trends of modern Russian economic system evidence a new wave of crisis of 2017. Therefore, at present (2018), entrepreneurial interests to maintain a favorable business climate (with a focus on the tax one) and social interests to prevent a sharp decline in the living standards step forward, but public interests related to the replenishment of all-level public budgets go into the background. That is, a scientific and practical issue of tax incentive for the growth and development of modern Russian social and economic system arises that is basic to crisis management.

In this regard, the microeconomic aspect of taxation connected with evaluation of tax burden on taxpayers and minimization of their tax payments, becomes especially urgent. Taking into account a big share of the shadow economy in modern Russia, we put forward a hypothesis that tax burden in Russia is high, and measures to minimize tax payments are limited and inefficient. The purpose of the paper is to study the essence of tax burden and mitigate tax payments and their peculiarities in modern Russia.

2 Materials and Method

Theoretical and methodological foundation of this study is the content of publications by modern authors on the topic of defining essence and measuring tax burden, among which Cushing and Newman (2018), Musayev and Musayeva (2018), Shi and Tao (2018), Strelnik et al. (2018), as well as matters of papers by various scientists on conceptual and applied issues of tax payments mitigation, including Gashenko et al. (2018), Katysheva (2016), Popkova et al. (2018a, b), Xu et al. (2015).

The methodological unit of this paper is built on a systematic approach and is based on the application of methods of structural and comparative analysis, synthesis, induction, deduction and formalization, as well as a complex of methods to estimate the value (in absolute terms) and the extent (in relative terms) of tax burden.

3 Results

In modern economic theory, tax burden denotes tax obligations of economic entities with the most important performance such as onerousness—a complex qualitative and quantitative evaluation by these entities of their tax conditions and their readiness to observe them. The bigger the tax burden, the more widespread the phenomena of tax opportunism in the economic system, i.e. tax evasion leading to an increase in the shadow economy extent.

From a qualitative point of view, tax burden is defined individually by each subject of taxation and is featured by a significant subjectivity, because it is confined to qualitative evaluation due to strong dependence on personal characteristics, peculiarities of their economic activities and other forces that cannot be measured accurately. From the standpoint of economic science, a qualitative evaluation of tax burden is technically complicated, therefore it is not carried out in this study.

From a quantitative point of view, tax burden is evaluated through the definition of extent and structure of aggregate tax revenues in the economy and their correlation to income of taxpayers—population and business. The results of evaluation of tax burden on the population and business in Russia as of the end of 2017-beginning of 2018 are presented in Table 1.

Table 1 Value and structure of tax burden on the population and business in Russia as of the end of 2017—beginning of 2018

Tax	Tax return, RUB billion	Share of population income, %	Share of company's sale revenues, %	Share of company profit, %
Mineral extraction tax	4130.4	–	2.77	35.65
Corporate income tax	3290	–	2.20	28.39
Property taxes (50%)	625.1	–	0.42	5.40
In total, corporate taxes	8045.5	–	**Tax burden on the business**	
			5.39	**69.44**
Personal income tax	3251.1	13.43	–	–
Social contributions	975.3	1.34	–	–
Value added tax	2069.9	8.55	–	–
Excise duties	1521.3	6.28	–	–
Property taxes (50%)	625.1	2.58	–	–
In total, personal and mixed taxes	8442.7	**Tax burden on the population 32.19**	–	–

Source Constructed and calculated by authors on the data (Federal Tax Service of the Russian Federation 2018; Rosstat 2018)

As can be seen from Table 1, we referred indirect taxes—value added tax and excise duties to the tax burden on the population, as far as they are consumption taxes and are paid actually by consumers. Social contributions are also designated as the tax burden on the population, despite the fact that their taxpayers are enterprises, because these contributions are paid from the payroll fund and lead to a decrease in salary amount.

The aggregate tax burden on the Russian population as of the end of 2017-beginning of 2018 is RUB 8442.7 billion, that is 32.19% of revenues (RUB 24,209.8 billion). In this structure the personal income tax (13.43% of the population income), value added tax (8.55% of the population income) and excise duties (6.28% of the population income) are prevailing. The share of property taxes on social contributions is 2.58 and 1.34% respectively. It should be noted that due to a lack of differentiation in collection of property taxes from the population and business for the purposes of this study, they are evenly distributed between these categories of taxpayers (50% by 50%).

The aggregate tax burden on the Russian business as of the end of 2017-beginning of 2018 is 5.39% of company sales revenue (RUB 149,334.2 billion) and 69.44% of company before-tax profit (RUB 11,587 billion). The biggest share belongs to the mineral extraction tax (2.77% of revenue and 35.65% of profit), followed by corporate income tax (2.20% of revenue and 28.39% of profit) and property taxes (0.42% of revenue and 5.40% of profit).

Let's us consider the tax burden on business in more detail by the example of the biggest Russian companies for capitalization at year-end 2017—PJSC "GAZPROM", PJSC "Sberbank of Russia" and PJSC OC "Rosneft", according to Russian Information Agency Rating (2018) (Table 2).

As can be seen from Table 1, the aggregate tax burden on PJSC "GAZPROM" as of the end of 2017-beginning of 2018 is 287.3 billion rubles (15% of sales revenue and 61% of before—tax profit), PJSC "Sberbank of Russia" is 287.3 billion rubles (12% of sales revenue and 33% of before-tax profit), PJSC OC "Rosneft" is

Table 2 Value and structure of the tax burden of the biggest Russian companies for capitalization at year-end 2017—PJSC "GAZPROM", PJSC "Sberbank of Russia" and PJSC OC "Rosneft" as of the end of 2017—beginning of 2018

Tax	PJSC "GAZPROM"			PJSC "Sberbank of Russia"			PJSC "Oil company Rosneft"		
	RUB billion	% of revenue	% of profit	RUB billion	% of revenue	% of profit	RUB billion	% of revenue	% of profit
Profit tax	57.2	3	13	250.4	11	29	98.0	2	25
Insurance contributions	9.0	0	2				61.0	1	15
VAT	39.1	2	9				185	3	47
Customs fees	16.0	1	4	0.1	0	0			
Other taxes	157.0	9	35	36.8	2	4			
Taxes in total	287.3	15	61	287.3	12	33	344.0	6	87

Source Constructed and calculated by authors on the data (PJSC GAZPROM 2018; PJSC Sberbank of Russia 2018; PJSC Rosneft Oil Company 2018)

RUB 344 billion (6% of sales revenue and 87% of before-tax profit). In the structure of tax burden, profit tax and mineral extraction tax classified as "other taxes" along with excise duties and property taxes are prevailing.

Consequently, the tax burden on business is mainly due to the peculiarities of doing business, where industry classification is primary. In general, the tax burden on both categories of taxpayers can be estimated quite high. It is documented by the data of economic freedom rating in the countries worldwide, where Russia ranked 45th in 2018 of 187 countries by the tax burden (85.8%) (Heritage Foundation 2018).

We can reduce legally the tax burden with the help of means to minimize tax payments. The peculiarities of their application in modern Russia are presented in Table 3.

As can be seen from Table 3, we pointed out three means to minimize tax payments. The first one is the arrangement of economic activity due to taxation. This method involves reducing the taxable base and refocusing of business activities at reduced tax rates.

The application of this mean by the population (natural entities) in modern Russia includes the restriction of property (voluntary refusal of partial assets) to reduce the taxable base for property taxes, because legal measures to decrease payments on personal income tax in Russia are not provided. Businesses (legal entities) have much more opportunities in application of this method, including selection of legal corporate form, scope of activity and accounting policy.

The second mean is the research and application of tax benefits. It means searching for legal measures to decrease tax payments while maintaining the taxable base. The application of this method by the population (natural entities) involves the re-registration of property on relatives to obtain benefits for pensioners on property taxes, as well as the acquisition of housing, education for yourself and children, treatment, etc. to get personal income tax deductions. Business (legal entities) can apply this method by choosing a tax treatment (benefits for priority entrepreneurial activities).

Table 3 Peculiarities of means to minimize tax payments in modern Russia by different categories of taxpayers

Mean to minimize tax payments	Peculiarities of application	
	By population (natural entities)	By business (legal entities)
Economic activity management due to taxation	Restriction of property (voluntary refusal of partial assets)	Selection of legal corporate form, scope of activity, accounting policy
Research and application of tax benefits	Re-registration of property, acquisition of housing, education, treatment	Tax treatment selection (benefits for priority types of entrepreneurship)
Reducing risk to apply tax sanctions	Research and strict observance of tax legislation	

Source Compiled by the authors

The third approach is decreasing the risk to apply tax sanctions (tax penalties, fines, etc.). The use of this method by both population (natural entities) and business (legal entities) means the research and strict observance of tax legislation.

4 Conclusion

Thus, the hypothesis under suggestion is proofed. We revealed that tax burden in modern Russia is a relatively high, which likely causes a shadow economy. The aggregated tax burden on the Russian population as of the end of 2017-beginning of 2018 amounts to 32%. It is mainly due to the personal income tax, the opportunities to be minimized are limited due to acquisition of tax benefits that requires significant time and resource expenditures (collecting, drawing up and certifying numerous supporting documents by a continually changing list and forms).

The aggregated tax burden on business in Russia as of the end of 2017-beginning of 2018 is 5.39% of sales revenues and 69.44% of before-tax profit. However, by the example of the biggest Russian companies for capitalization at year-end 2017—PJSC "GAZPROM", PJSC "Sberbank of Russia" and PJSC OC "Rosneft"—it is shown that it varies in practice from 6 to 15% of sales revenue and from 33 to 87% of before-tax profit. It is mainly due to the profit tax and the mineral extraction tax, with quite wide opportunities to minimize payments; and include both specialization in non-resource branches of economy and the selection of a special tax treatment.

Taking into account a revealed imbalance in tax burden distribution between the business and the population towards an increased burden on the latter, it is advised to provide in modern Russia additional opportunities to reduce tax burden on the population through simplifying procedure to obtain benefits on the profit tax payment.

References

Cushing, T. L., & Newman, D. (2018). Analysis of relative tax burden on nonindustrial private forest landowners in the Southeastern United States. *Journal of Forestry, 116*(3), 228–235.

Federal Service of State Statistics of the Russian Federation (Rosstat). (2018). Russia in Figures: A Brief Statistical Book. http://www.gks.ru/wps/wcm/connect/rosstat_main/rosstat/en/statistics/publications/catalog/doc_1135075100641. Data accessed: 10.06.2018.

Federal Tax Service of the Russian Federation. (2018). Tax Analytics: The Structure of Revenues into Consolidated Budget of the Russian Federation. https://analytic.nalog.ru/portal/index.ru-RU.htm. Data accessed: 10.06.2018.

Gashenko, I. V., Zima, Y. S., Stroiteleva, V. A., & Shiryaeva, N. M. (2018). The mechanism of optimization of the tax administration system. *Advances in Intelligent Systems and Computing, 622*, 291–297.

Katysheva, E. (2016). Methods of the taxation optimization for the oil-extracting companies in Russia. In *International Multidisciplinary Scientific GeoConference Surveying Geology and Mining Ecology Management, SGEM*, Vol. 3, pp. 357–364.

Musayev, A., & Musayeva, A. (2018). A study of the impact of the underground economy on the integral tax burden in the proportional growth model under uncertainty. *Advances in Fuzzy Systems, 2018*(630), 987.

PJSC "GAZPROM". (2018). Consolidated Interim Condensed Financial Statements Prepared in Accordance with International Financial Reporting Standards (IFRS) (not audited) as of March 31, 2018. http://www.gazprom.ru/f/posts/01/851439/gazprom-ifrs-1q2018-en.pdf. Data accessed: 10.06.2018.

PJSC "OC Rosneft". (2018). Consolidated Financial Statements as of December 31, 2017 with an independent auditor's opinion. https://www.rosneft.ru/upload/site1/document_cons_report/Rosneft_FS_12m2017_ENG.pdf. Data accessed: 10.06.2018.

PJSC "Sberbank of Russia". (2018). Consolidated Financial Statements over a year of 2017 with an independent auditor's opinion. http://www.sberbank.com/common/img/uploaded/files/info/Word_Rus_YE17–04fteet.pdf. Data accessed: 10.06.2018.

Popkova, E. G., Bogoviz, A. V., Lobova, S. V., & Romanova, T. F. (2018a). The essence of the processes of economic growth of socio-economic systems. *Studies in Systems, Decision and Control, 135,* 123–130.

Popkova, E. G., Bogoviz, A. V., Ragulina, Y. V., & Alekseev, A. N. (2018b). Perspective model of activation of economic growth in modern Russia. *Studies in Systems, Decision and Control, 135,* 171–177.

RIA Rating. (2018). The biggest Russian Companies for Capitalization—The 2017 Results. http://www.riarating.ru/infografika/20180130/630080911.html. Data accessed: 10.06.2018.

Shi, Y., & Tao, J. (2018). 'Faulty' fiscal illusion: Examining the relationship between revenue diversification and tax burden in major US cities across the economic cycle. *Local Government Studies, 2*(1), 1–20.

Strelnik, E. U., Usanova, D. S., Khairullin, I. G., Shafigullina, G. I., & Khairullina, K. T. (2018). Tax Burden in KPI system of corporation. *Journal of Engineering and Applied Sciences, 13*(2), 332–336.

The Heritage Foundation. (2018). Index of Economic Freedom. https://www.heritage.org/index/explore?view=by-variables. Data accessed: 10.06.2018.

Xu, Q., Wang, W., Xu, L., Fan, D. (2015). Apparel Supply Chain Optimization with Subsides under carbon emission taxation. In *2015 International Conference on Logistics, Informatics and Service Science, LISS*, p. 7369690.

Part III
Advanced Tools to Mitigate the Taxation System

Tax Holidays as an Upcoming Tool of Tax Incentive for Business Renewal

Sergey V. Shkodinsky, Alexander E. Suglobov, Oleg G. Karpovich, Olga V. Titova and Ekaterina A. Orlova

Abstract *Purpose* The purpose of the paper is to study modern Russian practice to grant tax holidays for entrepreneurship, substantiate the practicability of application as a tool of tax incentive for business renewal in modern Russia, reveal barriers and develop recommendations for its successful introduction in Russian tax practice. *Methodology* In the study we employed a complex of general scientific methods and the method of trend analysis. With their help we investigated trend data of Russian business renewal in 2007–2018. Information and analytical background of the study was a statistical data of Federal State Statistics Service of the Russian Federation (Rosstat), INSEAD, Cornell University, World Intellectual Property Organization and Federal Tax Service of the Russian Federation. *Results* It is revealed that since 2008 tax preferences have been applied for the purposes of tax incentive to upgrade entrepreneurship. However, they become a tax credit that is much less incentive than tax holidays. *Recommendations* To overcome the barrier

S. V. Shkodinsky (✉)
Moscow State Regional University, Moscow, Russia
e-mail: sh-serg@bk.ru

A. E. Suglobov
Financial University under the Government of the Russian Federation,
Moscow, Russia
e-mail: a_suglobov@mail.ru

O. G. Karpovich
Russian Customs Academy, Moscow, Russia
e-mail: iskran@yahoo.com

O. V. Titova
Altai State University, Barnaul, Russia
e-mail: otitova82@icloud.com

E. A. Orlova
State University of Management, Moscow, Russia
e-mail: e_a_orlovaguu@mail.ru

S. V. Shkodinsky
Research Institute of Finance of the Ministry of Finance of the Russian Federation,
Moscow, Russia

© Springer Nature Switzerland AG 2019
I. V. Gashenko et al. (eds.), *Optimization of the Taxation System: Preconditions, Tendencies, and Perspectives*, Studies in Systems, Decision and Control 182,
https://doi.org/10.1007/978-3-030-01514-5_8

on the way of tax holidays for tax incentive of Russian business renewal (related to tax evasion), we developed a proprietary algorithm.

Keywords Tax holidays · Tax incentive · Entrepreneurship renewal Modern Russia

JEL Classification E62 · H20 · K34

1 Introduction

In terms of the fourth industrial revolution and the transition of the world economic system to a new technological paradigm, the scientific and practical issue to upgrade world economies becomes more urgent. The Concept of Long-Term Social and Economic Development of the Russian Federation until 2020 approved by the decree N 1662-p. as of November 17, 2008 (Government of the Russian Federation 2018b) recognizes and highlights a necessary renewal of Russian entrepreneurship.

However, the practical implementation of business renewal in accordance with this concept is encumbered by the continuing imbalance and crisis of the Russian economy. So, due to one of the key performances of economic renewal—the global innovation index—Russia ranked 56th among countries worldwide (35.9 points) in 2011, and in 2013 it fell to 62nd, despite the increase in the value up to 37.2 points.

Although in 2017 the index increased to 38.7 points and the position in the global rating improved to 45th (INSEAD, Cornell University, World Intellectual Property Organization 2018), a new wave of economic crisis of 2017 may cause a further deterioration of positions in this rating. It evidences the necessity of additional public promotion to renew business in modern Russia.

The working hypothesis of this study is based on the assumption that tax holidays represent an advanced tool of tax incentive for entrepreneurship renewal. The purpose of the paper is to study modern Russian practice to grant tax holidays for entrepreneurship, substantiate the practicability of application as a tool of tax incentive for business renewal in modern Russia, reveal barriers and develop recommendations for its successful introduction in Russian tax practice.

2 Materials and Method

The essence of entrepreneurship renewal and peculiarities of implementation in modern Russia are investigated in the papers by scientists such as Bogoviz et al. (2018a, b) and Popkova et al. (2017, 2018). The conceptual fundamentals for granting tax holidays and the latest practices of their application are described in numerous publications by authors such as Bachek et al. (2012), Du et al. (2014) and Platikanova (2017).

The content analysis of above-mentioned publications showed that tax holidays are a tax preference associated with a temporary (over a particular tax period) provision of tax(es) relief to a taxpayer. We also found that the issues of applying tax holidays as a tool of tax incentive to renew business are almost not considered in available scientific studies and publications and are insufficiently investigated that was the reason for our study.

To define the opportunities and prospects for the application of tax holidays as a tool of tax incentive to renew business, we apply in this study a complex of general scientific methods and a trend analysis method to research trend data of Russian business renewal in 2007–2018 on Federal State Statistics Service of the Russian Federation (Rosstat) information.

3 Results

In modern Russia, tax holidays are granted only to small and medium business entities, newly registered for natural entities in the legal corporate form of individual enterprises that operate in specific economic branches under a simplified or patent tax treatment.

They were introduced by Federal Law No. 477-FZ dated December 29, 2014 "On Amendments to Part 2 of the Tax Code of the Russian Federation" and submitted to regional initiative (federal entities). They will be valid until 2020 and involve exemption from taxes on economic performance results for enterprises (alternatives to profit tax) for a term of no more than 2 calendar years (Government of the Russian Federation 2018c).

Tax holidays are intended to overcome the shadow business expansion in Russia and to renew industrial enterprises that are basic to modern Russian economy. Currently, a tax credit acts as a tool of tax incentive for business renewal, which is also affordable for big industrial enterprises. The conditions of tax credit provision are enshrined in Clause 67 of Tax Code of the Russian Federation (Part One) No. 146-FZ as of July 31, 1998 and in Decree of the Federal Tax Service of the Russian Federation No. MMB-7-8/683 as of December 16, 2016.

These conditions stipulate the provision of an investment tax credit for enterprises implementing renewal in the amount of up to 100% of investments (costs incurred to acquire the latest technologies and equipment and R&D) for a term of no more than 5 years, if available any supporting documents, a renewal business plan, bank guarantee and surety bond. The credit is provided for tax on profit and tax on property, which was renewed at a base interest rate of 0.5–0.75% of the Russian Federation Central Bank (Government of the Russian Federation 2018a; Federal Tax Service of the Russian Federation 2018).

Investment tax credit has been provided in Russia since 2008. To assess efficiency, let us turn to the trend data of Russian business renewal in 2007–2018 (data for 2017–2018 are projection), which is presented graphically in Fig. 1. These performances reflect R&D activity of enterprises—the number of R&D-performing

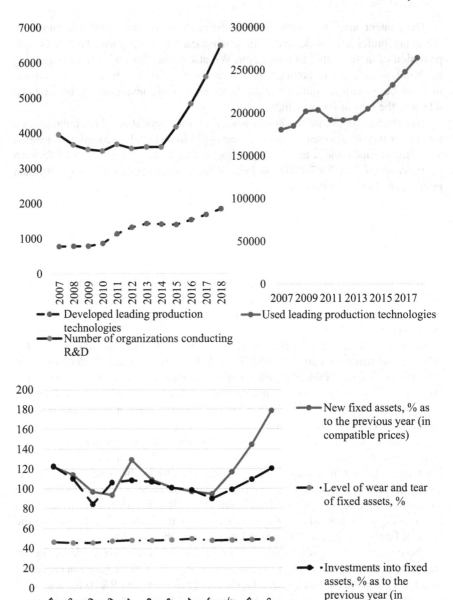

Fig. 1 Trend data of Russian business renewal in 2007–2018 (data for 2017–2018 are projection)
Source Compiled by the authors based on Rosstat (2018a, b)

companies, the number of developed and applied advanced manufacturing technologies, and also the intensity of fixed capital renewal—the degree of depreciation of fixed assets, new fixed capital formation, investments in fixed capital (fixed assets).

Figure 1 evidence that business activity in R&D and the rate of fixed capital renewal in Russia had been increased substantially during the tax credit action. Thus, the number of R&D-performing companies in 2017 against 2007 increased by 64.02%, the number of developed and applicable advanced manufacturing technologies—by 136.79 and 46.42% respectively. Despite an increase in the depreciation of fixed assets by 5.84% and a decrease in the annual growth rate of fixed capital investments by 1.87%, new fixed capital formation increased by 46.11%.

This stresses the practicability of granting tax preferences to enterprises undergoing renewal. At the same time, a far smaller increase in the intensity of fixed capital renewal in contrast to the growth of business activity in R&D field evidences a low tax credit attractiveness (due to the obligation to pay not only the entire amount of tax, but also interest thereon). A much more attractive alternative is the tax holidays (due to the lack of obligations to pay both interests, and the tax).

It should be noted that preference of tax holidays against a tax credit is recognized at government level, however, the choice toward credit is due to the high complexity (or even impossibility) of tax administration and control of tax holidays and protection of public tax interests. An analytical underpinning thereof was the possibility and high probability of tax credit evasion through bankruptcy (self-initiated, rather than influenced by objective market forces) of enterprises that took up a tax credit and their later registration as new legal entities.

Given pattern is not available for individual enterprises registered for natural entities and, therefore, featured by transparency and controllableness. To apply tax holidays as a tool of tax incentive for business renewal in modern Russia, we developed the following algorithm (Fig. 2), which helps to prevent tax credit evasion and grant large-scale availability, including big industrial enterprises.

As can be seen from Fig. 1, presented algorithm suggests consolidation of pubic property (the right of disposal) on innovation equipment and technologies acquired by the enterprise in terms of renewal, and the future results of R&D to grant tax preferences for.

Property (the right of disposal) will transfer to the enterprise in the event of a positive result of renewal (achievement of business plan goals, such as production of innovative products, launch of high-tech exports, increase of resource efficiency, development of advanced manufacturing technologies, etc.). At the same time, tax preferences become tax holidays.

If renewal has a negative result (business plan goals are not (fully) achieved), tax preferences take the form of a tax credit. After full implementation of obligation to pay the tax credit, the property (disposal of innovation equipment and technologies acquired by enterprise in terms of renewal and R&D results) will be transferred to enterprise. Otherwise (for example, if any violation of the tax credit repayment terms or business bankruptcy), it will be retained by state.

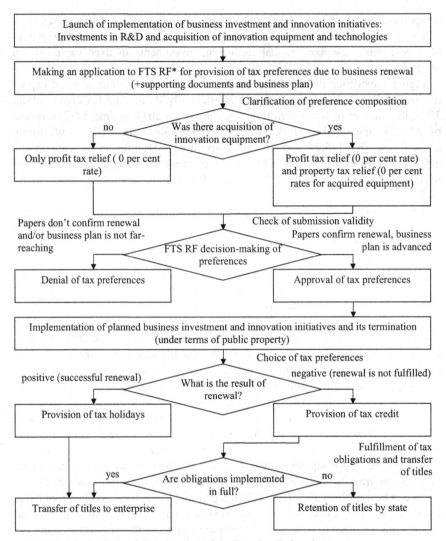

* FTS RF is the Federal Tax Service of the Russian Federation

Fig. 2 Advanced algorithm of tax incentive for Russian business renewal. *Source* Constructed by authors

Thereby, firstly, enterprises will be interested in actual renewal (targeted use of investments) and obtaining positive results and, secondly, they will not be interested in tax credit evasion, because in these circumstances they will suffer losses, and public tax interests will be protected.

4 Conclusion

In course of the study, we proofed the hypothesis under suggestion and proved that since 2008 tax preferences has been applied for the purposes of tax incentive for business renewal. However, they become a tax credit that is much less incentive than tax holidays. To overcome the barrier to tax holidays application for tax incentive of Russian business renewal (related to tax evasion), we developed a proprietary algorithm of tax incentive to renew Russian entrepreneurship.

The mentioned algorithm makes it possible to add positive result inducement, which will provide much more effective incentive, to the tax incentive of business renewal carried out on the basis of co-financing and crediting. It also allows to transfer risks of business renewal from the state to enterprises, since the form of granted tax preferences is defined due to results.

References

Bachek, Z. A., Ahmad, N., & Saleh, N. M. (2012). Correlation between tax holidays and earnings management: An empirical study. *Jurnal Pengurusan, 34,* 55–64.

Bogoviz, A. V., Ragulina, Y. V., Morozova, I. A., & Litvinova, T. N. (2018a). Experience of modern Russia in managing economic growth. In *Studies in systems, decision and control* (Vol. 135, pp. 147–154).

Bogoviz, A. V., Ragulina, Y. V., & Sirotkina, N. V. (2018b). Systemic contradictions in the development of modern Russia's industry in the conditions of establishment of the knowledge economy. In *Advances in intelligent systems and computing* (Vol. 622, pp. 597–602).

Du, L., Harrison, A., & Jefferson, G. (2014). FDI spillovers and industrial policy: The role of tariffs and tax holidays. *World Development, 64,* 366–383.

Federal Service of State Statistics of the Russian Federation (Rosstat). (2018a). *Russia in figures: A brief statistical book.* http://www.gks.ru/wps/wcm/connect/rosstat_main/rosstat/en/statistics/publications/catalog/doc_1135075100641 (data accessed: 11.06.2018).

Federal Service of State Statistics of the Russian Federation (Rosstat). (2018b). *Fixed assets.* http://www.gks.ru/wps/wcm/connect/rosstat_main/rosstat/en/statistics/enterprise/fund/# (data accessed: 11.06.2018).

Federal Tax Service of the Russian Federation. (2018). *Decree N MMB-7-8/683 of the Federal Tax Service of Russia as of December 16, 2016 "On approving the procedure for altering due date of taxes, fees, insurance contributions, as well as penalties and fines by tax authorities" (Registered in the Ministry of Justice of Russia on February 20, 2017 under No. 45707).* http://www.consultant.ru/document/Cons_doc_LAW_213100/08e747fee57b488ee702cf55dced9b76a3366b66/#dst100022 (data accessed: 11.06.2018).

Government of the Russian Federation. (2018a). *Tax code of the Russian Federation (part one) as of 31.07.1998 N 146-FZ: Clause 67. Procedure and conditions for granting an investment tax credit.* http://www.consultant.ru/document/cons_doc_LAW_19671/b27f4e6bdeea5735f6ef6e04fc6d4d9e649b0b40/ (data accessed: 11.06.2018).

Government of the Russian Federation. (2018b). *The concept of long-term social and economic development of the Russian Federation until 2020,* approved by the decree N 1662-p. as of November 17, 2008. http://www.consultant.ru/cons/cgi/online.cgi?req=doc&base=LAW&n=212832&fld=134&dst=100007,0&rnd=0.05004596916040649#05233422033651793 (data accessed: 11.06.2018).

Government of the Russian Federation. (2018c). *Federal Law as of December 29, 2014 N 477-FZ "On Amendments to Part 2 of the Tax Code of the Russian Federation"*. http://www.garant.ru/hotlaw/federal/592489/ (data accessed: 11.06.2018).

INSEAD, Cornell University, World Intellectual Property Organization (WIPO). (2018). *The global innovation index 2017*. http://www.wipo.int/edocs/pubdocs/en/wipo_pub_gii_2017.pdf (data accessed: 11.06.2018).

Platikanova, P. P. (2017). Investor-legislators: Tax holiday for politically connected firms. *British Accounting Review, 49*(4), 380–398.

Popkova, E. G., Popova, E. A., Denisova, I. P. P., & Porollo, E. V. (2017). New approaches to the modernization of the Russian and Greek regional economy. *European Research Studies Journal, 20*(1), 129–136.

Popkova, E. G., Bogoviz, A. V., Ragulina, Y. V., & Alekseev, A. N. (2018). Perspective model of activation of economic growth in modern Russia. In *Studies in systems, decision and control* (Vol. 135, pp. 171–177).

Taxation of Labor in Terms of Building a Social Market Economy

Z. V. Gornostaeva, I. V. Kushnareva and O. F. Sverchkova

Abstract *Purpose* The purpose of this paper is to reveal the features of non-conformity of labor taxation in modern Russia to the principles of social market economy and to develop the recommendations (measures) of improving labor taxation in terms of building a social market economy in Russia. *Methodology* In this study we apply the methods of system and problem analysis, synthesis, deduction, induction and formalization. The information and empirical background of research is modern (as of the end of 2017—beginning of 2018) statistical and analytical data of Federal State Statistics Service of Russia (Rosstat), the Institute of Social Sciences of the Russian Presidential Academy of the National Economy and Public Administration (RANEPA), the Government of the Russian Federation, and the United Nations Development Program. *Results* We revealed non-conformity of labor taxation in modern Russia to the principles of the social market economy, such as low public social guarantees in labor taxation and their complicated receipt, low level of social responsibility of employers and violation of labor taxation conditions, as well as poor contribution of labor taxation to establishing social justice. To improve labor taxation in modern Russia by bringing it into conformity with the principles of social market economy, we offered a complex of measures to overcome revealed unconformities: the expansion of public social guarantees within labor taxation and simplification of their receipt, tightening of administration of labor taxation and promotion of social responsibility of employers, as well as the introduction of a socially-oriented progressive taxation of labor. *Recommendations* We developed and advised a concept of labor taxation improvement in terms of building a social market economy for practical application in modern Russia.

Z. V. Gornostaeva (✉) · I. V. Kushnareva · O. F. Sverchkova
Institute of Service and Entrepreneurship (Branch) of Don State
Technical University, Shakhty, Russia
e-mail: zh.gornostaeva@mail.ru

I. V. Kushnareva
e-mail: innakusnareva@yandex.ru

O. F. Sverchkova
e-mail: ylo79@mail.ru

© Springer Nature Switzerland AG 2019
I. V. Gashenko et al. (eds.), *Optimization of the Taxation System: Preconditions, Tendencies, and Perspectives*, Studies in Systems, Decision and Control 182,
https://doi.org/10.1007/978-3-030-01514-5_9

Keywords Labor taxation · Social market economy · Taxation improvement
Modern Russia

JEL Classification E62 · H20 · K34

1 Introduction

The global economic crisis in the beginning of the 21st century foregrounded the
issue of building a social market economy, where the advantages of market freedom
such as high efficiency of economic activity and dynamic economic growth, are
supplemented by the advantages of a welfare state like social justice (redistribution
of population income in favor of the most vulnerable categories of the population),
social responsibility of employers and achievement of almost full engagement of
the population.

The working hypothesis of this study consists in that the labor taxation in
modern Russia does not fully conform to the principles of a social market economy.
The purpose of this paper is to reveal the features of non-conformity of labor
taxation in modern Russia to the principles of social market economy and to
develop the recommendations (measures) of improving labor taxation in terms of
building a social market economy in Russia.

2 Materials and Method

The theory and practice of labor taxation in modern economic systems is investigated in the papers by experts such as Anwar and Sun (2015), Bethencourt and
Kunze (2017), Churkin and Kalinina (2016), Fischer (2017), Gashenko et al.
(2018), Okunevičiūtė Neverauskienė et al. (2017), Sun and Anwar (2015). The
publications of scientists such as Bank (2017), Bogoviz et al. (2018a, b), Chabanet
(2017), Felice and Krienke (2017), Gallego-Álvarez and Quina-Custodio (2017),
Paraskewopoulos (2017), Popkova et al. (2017, 2018), Schlösser et al. (2017) are
devoted to the issues of building a social market economy.

At the same time, the peculiarities of labor taxation in terms of building a social
market economy are insufficiently explored and should be particularly investigated.
For this purpose, in this study we apply the methods of system and problem
analysis, synthesis, deduction, induction and formalization. The information and
empirical background of research is recent (as of the end of 2017—beginning of
2018) statistical and analytical data of Federal State Statistics Service of Russia
(Rosstat), the Institute of Social Sciences of the Russian Presidential Academy of
the National Economy and Public Administration (RANEPA), the Government of
the Russian Federation, and the United Nations Development Program.

3 Results

We have revealed the following features of non-conformity of labor taxation in modern Russia to the principles of social market economy as of the end of 2017— the beginning of 2018. Firstly, low social public guarantees and complicated receipt evidencing incomplete fulfillment of public obligations under labor taxation treaty. Despite a relatively high labor taxation (43% of salary per month), as a whole the level of human development in Russia is quite low in contrast to other developed countries—the value of the corresponding index is 0.798 points (ranked 50th of 188 countries) (United Nations Development Programme 2018).

Thus, the subsistence level in Russia in 2017 is fixed at 10,329 RUB per month (Government of the Russian Federation 2018). Insufficient amount for living in Russia underlines the provision of additional measures of social support for the least financially secure groups of population, such as subsidies for paying bills of housing and public services, etc. However, additional measures of social support are not provided immediately, but require a long (and regularly repeated) procedure of their multiple documentary confirmation and approval by different public authorities.

At the same time, unemployment benefits in Russia range from RUB 850–4900 per month (Rosstat 2018a), i.e. from 8.23 to 47.44% of the subsistence level, and mechanisms for bringing unemployed income to the subsistence level by state are not provided, that's why in Russia unemployed is a category of socially unprotected population.

The Russian average pension level (RUB 12,923 per month) exceeds the subsistence level for pensioners (RUB 8506 per month) by 51.93% (Rosstat 2018a). Nowadays, general retirement age is 55 years for women and 60 years for men. According to the Human Development Report, the average length of life in Russia is 70 years (United Nations Development Programme 2018). The current trend of increasing retirement age to 63 years for men and women implies an expected decline in living standards of the Russian population, as in this situation the retirement age (64–69 years) will be 6 years, i.e. 8.57% of life that is insignificant.

Secondly, low level of social responsibility of employers and their violation of labor taxation conditions. In 2017, 33 million people or 44.8% of the total employed population in Russia were involved in the shadow economy, i.e. they didn't have official job and didn't pay labor taxes. Taking into account that almost 70% of labor tax obligations (they are related to payment of social deductions from payroll fund) are imposed on employers, and then employees actually lack incentives for tax evasion.

On the contrary, if employers don't fulfill their labor tax obligations, then employees will not subsequently be able to claim necessary social guarantees from state, such as unemployment benefits and pensions (the minimum pension will be assessed). Applications by employees of shadow enterprises to the tax authorities about infringement of their tax rights by employers do not allow restoring social justice because legal entities in Russia (for example, limited liability companies—LLC) bear financial responsibility only in the amount of their charter capital (for LLC is, as a rule, RUB 10,000).

Therefore, low performance of labor tax administration combined with a low social responsibility of employers lead to violation of tax rights of employees in modern Russia.

Thirdly, labor taxation does not sufficiently contribute to the establishment of social justice. Russian current proportional system of taxation on personal income tax (13% of income—salary in the aspect of labor taxation) does not admit to redistribute incomes in society and lead to a gradual escalation of social stratification (differentiation of income of rich and poor). For comparison, in most modern developed countries (for example, in Germany) there is a progressive taxation scale that makes it possible to transfer the main burden to high-paid employees and reduce the tax burden on low-paid ones.

Announced public initiative to introduce a progressive scale of labor taxation in Russia is designed to solve the issue of the federal budget deficit and does not meet social interests. Thus, it is proposed to establish a reduced rate of personal income tax for the population with an annual income of less than 100 thousand RUB and to increase tax rate for the population with a higher annual income. This means that the tax rate will be decreased for social category of the population (practically non-existent) with income of less than RUB 8333.33 per month (80.68% of the subsistence level) and will be increased for the overwhelming majority of the population.

To improve labor taxation in modern Russia by bringing it into conformity with the principles of social market economy, we offer the following measures. The first measure is the introduction of a socially-oriented progressive scale of taxation. This measure is intended to overcome the federal budget deficit, provided that the level of social justice in Russia will be raised. We advise to establish the following tax rates of personal income tax for the most common labor categories in Russia:

– 10% rate for the population with income up to RUB 150,000 per year;
– 13% for the population with income of RUB 151,000–200,000 per year;
– 15% rate for the population with income of RUB 251,000–500,000 thousand per year;
– 20% rate for the population with income of RUB 501,000–1,000,000 per year;
– 25% rate for the population with income of over RUB 1,001,000 per year.

Tax consequences of the introduction of progressive taxation scale in Russia on the data for 2017 are evaluated in Table 1.

As is seen from Table 1, it is possible to enumerate 5 categories of the population according to the income level in Russia. The introduction of progressive scale of taxation will allow to decrease the tax burden on the category with the lowest income (from RUB 389.37 billion to RUB 299.52 billion), maintain current tax burden for the population with an average income (151,000–200,000 per year), increase it slightly for the population with above-average income (RUB 251,000–1,000,000 per year) and increase significantly for the population with a very high

Table 1 Evaluation of tax consequences of the introduction of progressive taxation scale in Russia on the data for 2017

Category by income level	Share of category in revenue structure of the population (%)	Total income of category (RUB billion)	Tax payments at 13% rate (RUB billion)	Tax payments under progressive scale (RUB billion)
Up to RUB 150,000 per year	5.4	2995.196	389.3755	299.5196
RUB 151,000–250,000 per year	10.1	5602.127	728.2765	728.2765
RUB 251,000–500,000 per year	15.1	8375.457	1088.809	1256.318
RUB 251,000–500,000 per year	22.6	12,535.45	1629.609	2507.09
over RUB 1001,000 per year	46.8	25,958.37	3374.588	6489.592
In total	100.0	55,466.6	7210.6 (2017)	11,280.8 (+17.76%)

Source Calculated by authors on the data (Rosstat 2018b)

income (from RUB 3374.58 billion to RUB 6489.59 billion). And thus, the aggregate labor tax revenues to the budget will increase by 17.76%, and social justice will rise.

The second measure is tightening of labor tax administration and promoting social responsibility of employers. It is necessary to introduce a system of obligatory insurance of enterprises if any violations of labor taxation conditions take place. This will restore social justice in the event of revealing tax evasion by employers.

The third measure is the expansion of public social guarantees in labor taxation and simplification of receipt procedure. In these terms, we advise to resume Russian practice of diversifying pension assets and providing employees with the opportunity to dispose of pay component of their pension. It is also proposed to replace subsidies for low-paid population with tax benefits on personal income tax (a reduced rate of 8% is recommended), which will be provided automatically on the tax service data.

Given recommendations are depicted in the concept to improve labor taxation in terms of building a social market economy in Russia (Fig. 1).

As can be seen from Fig. 1, the background of improving labor taxation is the introduction of a socially-oriented progressive scale of taxation that allows

Fig. 1 The concept to improve labor taxation in terms of building a social market economy in Russia. *Source* Drawn up by authors

overcoming the deficit of tax revenues and provide an opportunity for full-scale observance of public social guarantees of labor taxation. The central core is tightening of labor tax administration and promotion of social responsibility of employers to raise tax protection of employees.

The top is the expansion of public social guarantees within labor taxation and simplification of receipt procedure, ensuring adjustment of public and social interests in labor taxation and tax incentive of employment. As a result, we reach purpose of bringing labor taxation of modern Russia in conformity with the principles of social market economy and ensure the growth of employment rate, overcome shady employment of the population, and the growth of social security of the population in modern Russia.

4   Conclusion

Summing up the conducted research, it is possible to draw a conclusion that the hypothesis under suggestion is correct. We have revealed non-conformity of labor taxation in modern Russia to the principles of the social market economy, such as low public social guarantees in the field of labor taxation and complicated receipt, low level of social responsibility of employers and violation of labor taxation conditions, as well as poor contribution of labor taxation to establishing social justice in society.

To improve labor taxation in modern Russia by bringing into conformity with the principles of social market economy, we have offered a complex of measures to overcome revealed unconformities: the expansion of public social guarantees within labor taxation and simplification of their receipt, tightening of labor tax administration and promotion of social responsibility of employers, as well as the introduction of a socially-oriented progressive labor taxation. We have developed and advised a concept of labor taxation improvement in terms of building a social market economy for practical application in modern Russia.

References

Anwar, S., & Sun, S. (2015). Taxation of labour income and the skilled–unskilled wage inequality. *Economic Modelling, 47*, 18–22.

Bank, D. (2017). The double-dependent market economy and corporate social responsibility in Hungary. *Corvinus Journal of Sociology and Social Policy, 8*(1), 25–47.

Bethencourt, C., & Kunze, L. (2017). Temptation and the efficient taxation of education and labor. *Metroeconomica, 68*(4), 986–1000.

Bogoviz, A. V., Ragulina, Y. V., Morozova, I. A., & Litvinova, T. N. (2018a). Experience of modern Russia in managing economic growth. In *Studies in systems, decision and control* (Vol. 135, pp. 147–154).

Bogoviz, A. V., Ragulina, Y. V., & Sirotkina, N. V. (2018b). Systemic contradictions in development of modern Russia's industry in the conditions of establishment of knowledge economy. In *Advances in intelligent systems and computing* (Vol. 622, pp. 597–602).

Chabanet, D. (2017). The social economy sector and the welfare state in France: Toward a takeover of the market? *VOLUNTAS: International Journal of Voluntary and Nonprofit Organizations, 28*(6), 2360–2382.

Churkin, V. I., & Kalinina, O. V. (2016). Estimation of excess burden of labor taxation in Russia. *Actual Problems of Economics, 184*(10), 278–282.

Federal Service of State Statistics of the Russian Federation (Rosstat). (2018a). *Social and economic situation in Russia of 2017.* http://www.gks.ru/free_doc/doc_2017/social/osn-06-2017.pdf (data accessed: 12.06.2018).

Federal Service of State Statistics of the Russian Federation (Rosstat). (2018b). *Standard of living: Scope and structure of population income due to source of receipt; distribution of the total income and features of differentiation of population incomes.* http://www.gks.ru/wps/wcm/connect/rosstat_main/rosstat/en/statistics/population/level/# (data accessed: 12.06.2018).

Felice, F., & Krienke, M. (2017). Understanding social market economy, Francesco Forte and his interpretation. *International Advances in Economic Research, 23*(1), 21–37.

Fischer, T. (2017). Can redistribution by means of a progressive labor income-taxation transfer system increase financial stability? *JASSS, 20*(2), 3–5.

Gallego-Álvarez, I., & Quina-Custodio, I. A. (2017). Corporate social responsibility reporting and varieties of capitalism: An international analysis of state-led and liberal market economies. *Corporate Social Responsibility and Environmental Management, 24*(6), 478–495.

Gashenko, I. V., Zima, Y. S., Stroiteleva, V. A., & Shiryaeva, N. M. (2018). The mechanism of optimization of the tax administration system with the help of the new information and communication technologies. In *Advances in intelligent systems and computing* (Vol. 622, pp. 291–297).

Government of the Russian Federation. (2018). *Resolution N 1119 as of September 19, 2017 "On establishing subsistence level per capita and for the main social and demographic groups of the population across the Russian Federation for the second quarter of 2017".* http://www. consultant.ru/document/cons_doc_LAW_278382/ (data accessed: 12.06.2018).

Okunevičiūtė Neverauskienė, L., Miežienė, R., & Gataūlinas, A. (2017). Evaluation of the relationship between labour taxation and unemployment: Case study of Lithuania in the EU context | [Darbo apmokestinimo ir nedarbo saryšio vertinimas: Lietuvos atvejo analize ES kontekste]. *Filosofija, Sociologija, 28*(4), 225–235.

Paraskewopoulos, S. (2017). The German model of "Social Market Economy". In *Democracy and an open-economy world order*, (pp. 83–91). Springer International Publishing.

Popkova, E. G., Popova, E. A., Denisova, I. P., & Porollo, E. V. (2017). New approaches to modernization of spatial and sectorial development of Russian and Greek regional economy. *European Research Studies Journal, 20*(1), 129–136.

Popkova, E. G., Bogoviz, A. V., Ragulina, Y. V., & Alekseev, A. N. (2018). Perspective model of activation of economic growth in modern Russia. In *Studies in systems, decision and control* (Vol. 135, pp. 171–177).

Schlösser, H. J., Schuhen, M., & Schürkmann, S. (2017). The acceptance of the social market economy in Germany. *Citizenship, Social and Economics Education, 16*(1), 3–18.

Sun, S., & Anwar, S. (2015). Taxation of labour, product varieties and skilled–unskilled wage inequality: Short run versus long run. *International Review of Economics and Finance, 38,* 250–257.

United Nations Development Programme. (2018). Human Development Report 2016. http://hdr. undp.org/en/countries/profiles/RUS (data accessed: 12.06.2018).

Restructuring of Tax Liabilities as an Upcoming Trend of Economic Diversification in Modern Russia

Natalya A. Shibaeva, Anna I. Zarudneva, Anastasia A. Sozinova, Alexander V. Shuvaev and Alexander N. Alekseev

Abstract *Purpose* The purpose of the paper is to substantiate possible and necessary use of restructuring of tax liabilities as an upcoming trend of economic diversification in modern Russia, and to develop practical recommendations. *Methodology* In this paper we apply a complex of general scientific methods, as well as methods of time-series analysis in economic statistics—horizontal and trend analysis. Information and empirical background is provided by data of the Center for Macroeconomic Analysis and Short-Term Forecasting of the Russian Federation, RIA Rating, and Federal State Statistics Service of the Russian Federation (Rosstat). To define the relevance of the mechanism for restructuring tax liabilities in modern Russia, the authors address to dynamic pattern of enterprises and closed down ones due to bankruptcy in 2007–2017. *Results* As a result, we had proved that the restructuring of tax liabilities is an upcoming trend of economic diversification in modern Russia. It is revealed that over recent years, the demand for this mechanism in relation to growing unprofitability of enterprises and their frequent close-down due to bankruptcy has increased in Russian economic system. However, the state introduces tight restrictions for enterprises restructuring their tax liabilities to protect own interests (guaranteed replenishment of budget and

N. A. Shibaeva (✉)
Orel State University, Orel, Russia
e-mail: super-ya-57@mail.ru

A. I. Zarudneva
Volgograd State Technical University, Volgograd, Russian Federation
e-mail: volgoecona@yandex.ru

A. A. Sozinova
Vyatka State University, Kirov, Russian Federation
e-mail: 1982nastya1982@mail.ru

A. V. Shuvaev
Stavropol State Agrarian University, Stavropol, Russian Federation
e-mail: a-v-s-s@rambler.ru

A. N. Alekseev
Financial University Under the Government of the Russian Federation, Moscow, Russia
e-mail: Alexeev_alexan@mail.ru

© Springer Nature Switzerland AG 2019
I. V. Gashenko et al. (eds.), *Optimization of the Taxation System: Preconditions, Tendencies, and Perspectives*, Studies in Systems, Decision and Control 182,
https://doi.org/10.1007/978-3-030-01514-5_10

extra-budgetary funds). *Recommendations* In order to ensure widespread avail-ability of the mechanism to restructure tax liabilities for modern Russian enter-prises, it is advised legislative consolidation of priority implementation of restructured tax liabilities. To aim this mechanism at diversification of the Russian economy, we propose to grant preferential terms (reduced limit of tax liabilities and extended term of their restructuring) for enterprises of the most promising and strategically crucial branches of national economy in terms of economic diversification.

Keywords Restructuring of tax liabilities · Economic diversification
Modern Russia

JEL Classification E62 · H20 · K34

1 Introduction

One of the most important consequences of the global economic recession of recent years was large-scale bankruptcy of enterprises around the world. Industrial enterprises, especially mining industry, have been the most resistant to crisis. Countries with a developed and prevailing industry of the sectoral composition, such as China, Japan, etc., had been least affected by negative impact of external crisis forces and had recovered more quickly.

Industrial orientation is a characteristic feature of the Russian modern economy. As of 2018, the total share of mining and manufacturing industries in the sectoral composition of Russian GDP is 32.7%, and in the composition of Russian exports is 77.6% (Rosstat 2018). On the one hand, developed industrial sector allowed mitigating the effect of a global recession in the Russian economy. On the other hand, the industrial specialty of the economy causes the dependence of modern Russia on the import of goods from other sectors of the national economy that exacerbates the issue of ensuring national economic security (in context of inde-pendence and self-sufficiency of commercial products) in terms of the world eco-nomic crisis.

In this regard, Concept of long-term social and economic development of the Russian Federation until 2020 approved by the decree N 1662-p. as of November 17, 2008 emphasizes the importance of economic diversification in modern Russia with focus on trade, agriculture and advanced technologies (Government of the Russian Federation 2018a). However, its diversification is encumbered because of ongoing economic crisis. The working hypothesis of this study consists in the assumption that an upcoming trend of economic diversification in modern Russia is the restructuring of corporate tax liabilities. The purpose of the paper is to sub-stantiate possible and necessary use of restructuring of tax liabilities as an upcoming trend of economic diversification in modern Russia, and to develop practical recommendations.

2 Materials and Method

Theory and practice of restructuring corporate tax liabilities and its role in public promotion of business development is discussed in the papers by scholars such as Cen et al. (2017), Inamura and Okuda (2017), Isin (2018), Norbäck et al. (2018), Paes (2014), Shin and Woo (2018). The advantages and disadvantages of strategies for industrial specialty and economic diversification are analyzed in the publications by scientists such as Akram and Rath (2017), Clark et al. (2017), Howie (2018), Lefèvre (2017), Popkova et al. (2018a, b, c).

At the same time, the prospects to use the restructuring of corporate tax liabilities for benefits of economic diversification remain out of view by authors of most available publications. To study them, we applied a complex of general scientific methods (analysis, synthesis, induction, deduction and formalization), as well as methods of time-series analysis in economic statistics-horizontal and trend analysis.

Information and empirical background is provided by data of the Center for Macroeconomic Analysis and Short-Term Forecasting of the Russian Federation, RIA Rating, and the Federal State Statistics Service of the Russian Federation (Rosstat). To define the relevance of the mechanism for restructuring tax liabilities in modern Russia, the authors address to dynamic pattern of enterprises and closed down ones due to bankruptcy of 2007-2017 that depicted in Fig. 1.

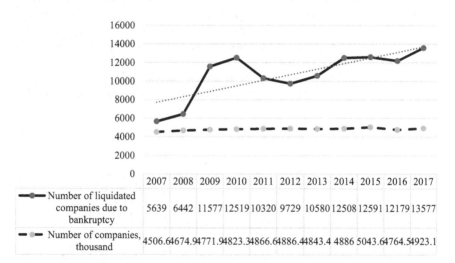

Fig. 1 Dynamic pattern of Russian enterprises and ones closed down due to bankruptcy in 2007–2017. *Source* Compiled by the authors based on Center for Macroeconomic Analysis and Short-Term Forecasting (2018), Rosstat (2018)

3 Results

Analysis of the data in Fig. 1 showed that the number of Russian enterprises had been quite stable throughout 2007–2017 and hadn't had negative dynamics even within Russian economy crisis in 2009–2010; its ten-year trend (growth in 2017 against 2007) was 9.24%. The number of enterprises closed down due to bankruptcy had sharply increased during the crisis from 6442 units in 2008 to 11,577 units in 2009 and 12,519 units in 2010.

This ten-year trend is still positive (amounts to 140.77%). It means that the number of Russian enterprises closed down due to bankruptcy has increased almost 1.5-fold over the past 10 years. This share of the total number of enterprises also has increased significantly during the crisis, from 0.14% in 2008 to 0.24% in 2009 and 0.26% in 2010. It is notable that it repeatedly increased in 2014 to 0.26%, and in 2017 to 0.28%. Its ten-year trend is 120.4%.

According to RIA Rating data, the share of Russian unprofitable enterprises in 2017 is 29.39% (RIA Rating 2018). It means that almost a third of Russian enterprises are teetering on the edge of bankruptcy. Consequently, the need (potential demand) of modern Russian enterprises for restructuring of tax liabilities is enough high; it has grown annually, especially in terms of crisis [the last has taken place in the Russian economy today (2017–2018)].

The restructuring of tax liabilities is a process to alter the term structure and procedure for paying taxes and fees to public budgets and extra-budgetary funds. This implies that upon receipt of the right to restructure tax liabilities, an enterprise may pay them later under strictly agreed schedule approved by the tax service instead of a lump-sum payment of taxes and fees within legal period. The logical pattern for restructuring tax liabilities is presented by us in Fig. 2.

As you can see in the Fig. 2, the restructuring of tax liabilities is applied in terms of enterprise crisis. The reasons may be different and covering violation of internal/

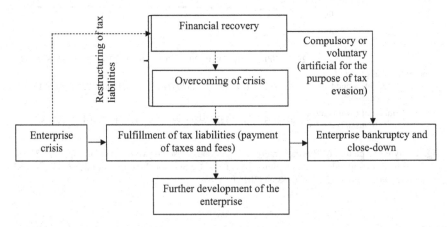

Fig. 2 Logical pattern for restructuring tax liabilities. *Source* Compiled by the authors

external economic relations and contract terms by business partners, a sharp decline in demand for goods, an unsuccessful performance of innovative activity by the enterprise, and many others. In the event of a crisis, if an enterprise fulfills its tax liabilities (it will pay all taxes and fees in full and in due time), it will be closed down due to bankruptcy.

To avoid this, it is applied the restructuring of tax liabilities which allows to initiate financial recovery of an enterprise and overcome crisis. After that, the company will be able to fulfill its previous and current tax liabilities (to pay taxes and fees) and develop further. Restructuring of tax liabilities always bears a risk of business bankruptcy in the event of unsuccessful financial recovery. The bankruptcy of an enterprise with restructured tax liabilities may be compulsory (due to objective market reasons) or voluntary (artificial for the purpose of tax evasion).

The government contemplates strict rules and restrictions on the provision of possible restructuring of tax liabilities to prevent tax evasion by enterprises. According to Clause 64 "The procedure and terms for granting deferrals or installments to pay taxes, fees, insurance contributions" of Tax Code of the Russian Federation (part one) as of July 31, 1998 N 146-FZ, the restructuring of tax liabilities in Russia is provided on the following terms (Government of the Russian Federation 2018b):

- Deferrals or installments (tax credit) of taxes and fees with the total amount of no less than RUB 100,000 for a term of 1–3 years;
- If the financial standing of the company does not allow to pay taxes and fees in due time, but it will be highly likely improved within the term of deferral or installment provision;
- The term of incorporation is not less than a year ago; the company is not in reorganization or close-down; there is no any opened criminal investigation.

We propose the following practical recommendations to use the restructuring of tax liabilities as an upcoming trend of economic diversification in modern Russia. Firstly, it is advised to expand the application of restructuring tax liabilities in the most priority sectors of the national economy designed to diversify the Russian economy.

These are agro-industry, where the development of entrepreneurship is strategically crucial to ensure national food security, and industry of advanced technologies, where the development of entrepreneurship is necessary to ensure the global long-term competitiveness of the economy.

For these industries, it is meaningful to reduce the amount of tax liabilities (up to 50 thousand rubles) and increase the term of restructuring (up to 5 years).

Secondly, it is necessary to enshrine in law the priority payment of restructured tax liabilities. In terms of building a social market economy, a constraint to the widespread provision of opportunities to restructure tax liabilities is not only the risk of a deficit in all-level government budgets, but also the risk of a deficit in public extra-budgetary funds paying social security and pension benefits to the population.

Therefore, in the interest of using the restructuring of tax liabilities for economic diversification, which requires its widespread provision, it is necessary to define priority payment of taxes and fees as follows: first of all, social contributions; then federal taxes and fees (value added tax, excise duties, mineral extraction tax and corporate profit tax) and finally territorial taxes and fees (property taxes).

Practical implementation of mentioned recommendations will help to reduce risks and thereby overcome barriers to widespread provision of the opportunity to restructure tax liabilities for modern Russian enterprises. The provision of preferential conditions to restructure tax liabilities for enterprises of priority sectors of the national economy allows both to promote the Russian economy diversification, and to ensure its economic security and global long-term competitiveness.

4 Conclusion

Thus, a developed hypothesis had been proofed. We had proved that the restructuring of tax liabilities is an upcoming trend of economic diversification in modern Russia. It is revealed that over recent years, the demand for this mechanism in relation to growing unprofitability of enterprises and their frequent close-down due to bankruptcy has increased in Russian economic system. However, the state introduces tight restrictions for enterprises restructuring their tax liabilities to protect own interests (guaranteed replenishment of budget and extra-budgetary funds).

In order to ensure widespread availability of the mechanism to restructure tax liabilities for modern Russian enterprises, it is advised legislative consolidation of priority implementation of restructured tax liabilities. To aim this mechanism at diversification of the Russian economy, we propose to grant preferential terms (reduced limit of tax liabilities and extended term of their restructuring) for enterprises of the most promising and strategically crucial branches of national economy in terms of economic diversification.

References

Akram, V., & Rath, B. N. (2017). Export diversification and sources of growth in emerging market economies. *Global Economy Journal, 17*(3), 20170018.

Cen, W., Tong, N., & Sun, Y. (2017). Tax avoidance and cost of debt: Evidence from a natural experiment in China. *Accounting and Finance, 57*(5), 1517–1556.

Center for Macroeconomic Analysis and Short-Term Forecasting. (2018). Bankruptcy of legal entities in Russia: the main trends. http://www.forecast.com/_ARCHIVE/Analitics/PROM/2017/Bnkrpc-4-16_v3.pdf. Data accessed: 12.06.2018.

Clark, D. P. P., Lima, L. R., & Sawyer, W. C. (2017). Stages of diversification in high performing Asian economies. *Journal of Economic Studies, 44*(6), 1017–1029.

Federal Service of State Statistics of the Russian Federation (Rosstat). (2018). Russia in figures: A brief statistical book. http://www.gks.ru/wps/wcm/connect/rosstat_main/rosstat/en/statistics/publications/catalog/doc_1135075100641. Data accessed: 12.06.2018.

Government of the Russian Federation. (2018a). The concept of long-term social and economic development of the Russian Federation until 2020 approved by the decree No. 1662-r dated 17, 2008.   http://www.consultant.ru/cons/cgi/online.cgi?req=doc&base=LAW&n=212832&fld= 134&dst=100007,0&rnd=0.05004596916040649#05233422033651793. Data accessed: 12. 06.2018.

Government of the Russian Federation. (2018b). Tax code of the Russian Federation (Part One) as of July 31, 1998 No. 146-FZ: Clause 64. Procedure and terms for granting deferrals or installments to pay taxes, fees, insurance contributions. http://www.consultant.ru/document/ cons_doc_LAW_19671/e7c6a6a5f85706a084493bb1c21b3e97932d003d/. Data accessed: 13. 06.2018.

Howie, P. P. (2018). Policy transfer and diversification in resource-dependent economies: Lessons for Kazakhstan from Alberta. *Politics and Policy, 46*(1), 110–140.

Inamura, Y., & Okuda, S. (2017). Deferred taxes and cost of debt: Evidence from Japan. *Asia-Pacific Journal of Accounting and Economics, 24*(3–4), 358–376.

Isin, A. A. (2018). Tax avoidance and cost of debt: The case for loan-specific risk mitigation and public debt financing. *Journal of Corporate Finance, 49,* 344–378.

Lefèvre, R. (2017). The Algerian economy from 'oil curse' to 'diversification'? *Journal of North African Studies, 22*(2), 177–181.

Norbäck, P. P.-J., Persson, L., & Tåg, J. (2018). Does the debt tax shield distort ownership efficiency? *International Review of Economics and Finance, 54,* 299–310.

Paes, N. L. (2014). The effects of installments on tax collection|[Os efeitos dos parcelamentos sobre a arrecadação tributária]. *Estudos Economicos, 44*(2), 323–350.

Popkova, E. G., Bogoviz, A. V., Lobova, S. V., & Romanova, T. F. (2018a). The essence of the processes of economic growth of socio-economic systems. *Studies in Systems, Decision and Control, 135,* 123–130.

Popkova, E. G., Bogoviz, A. V., Pozdnyakova, U. A., & Przhedetskaya, N. V. (2018b). Specifics of economic growth of developing countries. *Studies in Systems, Decision and Control, 135,* 139–146.

Popkova, E. G., Bogoviz, A. V., Ragulina, Y. V., & Alekseev, A. N. (2018c). Perspective model of activation of economic growth in modern Russia. *Studies in Systems, Decision and Control, 135,* 171–177.

RIA Rating.(2018). Regional social and economic rating of 2017. http://www.riarating.ru/ infografika/20170530/630063754.html. Data accessed: 12.06.2018.

Shin, H.-J., & Woo, Y.-S. (2018). The effect of tax avoidance on the value of debt capital: Evidence from Korea. *South African Journal of Business Management, 48*(4), 83–89.

Effective Tax Policy of the State: Conceptual Foundations and Methodology of Evaluation

Aleksei V. Bogoviz, Tatyana A. Zhuravleva, Elena G. Popkova, Anna I. Zarudneva and Marina L. Alpidovskaya

Abstract *Purpose* The purpose of the research is to study conceptual foundations and methodology of evaluation of effectiveness of state tax policy and to determine effectiveness of state tax policy that is implemented in modern Russia. *Methodology* For complex evaluation of effectiveness of state tax policy, the proprietary methods, which allows combining advantages of both existing conceptual approaches and overcoming their drawbacks, is used. This method envisages evaluation of effectiveness of state tax policy through separate calculation of the value of financial indicator and values of non-financial indicators with the following unification and treatment of the received results. *Results* It is showed that the modern Russian state tax policy is peculiar for low effectiveness. The main reason for that is insufficiently successful implementation of the most important function of the taxation system— provision of collection of taxes for state budgets of all levels of the budget system— due to deficit of the consolidated state budget of the Russian Federation and critically large volume of tax evasion (shadow economy). Costs of tax policy exceed its positive results by more than two times, even without consideration of expenditures

A. V. Bogoviz (✉)
Federal State Budgetary Scientific Institution "Federal Research Center of Agrarian
Economy and Social Development of Rural Areas—All Russian
Research Institute of Agricultural Economics", Moscow, Russia
e-mail: aleksei.bogoviz@gmail.com

T. A. Zhuravleva
Orel State University, Orel, Russia
e-mail: orel-osu@mail.ru

E. G. Popkova
Institute of Scientific Communications, Volgograd, Russia
e-mail: 210471@mail.ru

A. I. Zarudneva
Volgograd State Technical University, Volgograd, Russian Federation
e-mail: volgoecona@yandex.ru

M. L. Alpidovskaya
Financial University Under the Government of the Russian Federation,
Moscow, Russian Federation
e-mail: morskaya67@bk.ru

© Springer Nature Switzerland AG 2019
I. V. Gashenko et al. (eds.), *Optimization of the Taxation System: Preconditions,
Tendencies, and Perspectives*, Studies in Systems, Decision and Control 182,
https://doi.org/10.1007/978-3-030-01514-5_11

for state tax administration and control. In addition to this, state tax policy in Russia does not fully conform to the declared principles of stability, transparency, justice, and stimulation of national interests. *Recommendations* it is concluded that low effectiveness of tax policy of the state could be one of the reasons of non-optimality of modern Russia's taxation system. That's why increase of effectiveness of state tax policy is recommended for optimization of this system.

Keywords Effective tax policy · Taxation system · Modern Russia

JEL Classification E62 · H20 · K34

1 Introduction

Effectiveness is the most important characteristic of all economic phenomena and processes, as the least effective of them are inexpedient and should be terminated and deleted for successful functioning and development of economic systems. The necessity for practice of taxation is beyond any doubt and is caused by the necessity for redistribution of revenues in society; however, successfulness of its implementation could be different depending on the implemented state tax policy. It is a process of managing the tax system by determining the structure and characteristics of taxes, as well as provision, monitoring, and control of their collection.

Effectiveness of state tax policy determines the work of the taxation system and, therefore, ensures possibilities for replenishment of state budgets of all levels of the budget system and sets tax climate for economic subjects. That's why tracking the effectiveness of state tax policy is an important task for modern economic science and practice, as it allows determining its current level, comparing it to its previous periods, and forecasting its changes in the future, as well as conducting international comparisons of economic systems as to the level of effectiveness of state tax policy.

Evaluation of effectiveness of state tax policy becomes especially topical in the context of optimization of the taxation system. Based on the determined signs of non-optimality of the modern Russia's taxation system, we offer a hypothesis that tax policy of Russia is peculiar for low effectiveness. The purpose of the research is to study conceptual foundations and methodology of evaluation of effectiveness of state tax policy and to determine effectiveness of state tax policy that is implemented in modern Russia.

2 Materials and Method

From the scientific and theoretical point of view, effectiveness of state tax policy—
as well as any other economic process—should be evaluated through comparison of
expenditures and results; the larger this ratio, the higher the effectiveness.
According to this, from the practical point of view, there are two co conceptual
approaches to determining the effectiveness of state tax policy.

The first approach—functional—envisages evaluation of effectiveness of tax
policy through the prism of its financial component, determining result as successfulness of taxation system's execution of its income function (replenishment of
state budget), and expenditures as costs of this result—expenditures for tax
administration and control. This approach is described in detail in the works:
Condie et al. (2017), Ferré et al. (2018), Lewis (2018), Muennig et al. (2016),
Popkova et al. (2018a, b), Yanıkkaya and Turan (2018).

The advantage of this approach is foundation on precise quantitative indicators,
which guarantees objectivity and compatibility of results. The drawback is consideration of only financial characteristics of state tax policy and neglecting its other
characteristics.

This drawback could be overcome within the second approach—essential—
which envisages evaluation of effectiveness of tax policy through the prism of its
correspondence to the principles of taxation, determining the result as successfulness and complete correspondence of the taxation system to its principles, and
expenditures as lack of correspondence to these principles, i.e., violations in the
work of the taxation system. Variations and interpretations of this approach are
presented in the works: Chalu and Mzee (2018), Crespi et al. (2016), Gashenko
et al. (2018), Jang and Eger (2018), Thomson (2017).

The advantage of this approach is detailed evaluation and consideration of the
fullest specter of indicators of effectiveness of state tax policy. At that, its serious
drawback is related to foundation on qualitative indicators, due to which the results
are subjective and are, as a rule, imprecise.

3 Results

For complex evaluation of effectiveness of state tax policy we developed a proprietary method that allows combining advantages of both existing conceptual
approaches and overcoming their drawbacks. The offered method envisages conduct of evaluation of effectiveness of state tax policy via separate calculation of the
value of the financial indicator and values of non-financial indicators with further
unification and general treatment of obtained results.

Calculation of the financial indicator of effectiveness of state tax policy within
the developed method is performed with the following formula:

$$FIestp = (TRCSB * (RCSB)/(SETAC + VTL), \qquad (1)$$

where

FIestp—financial indicator of effectiveness of state tax policy;

TRCSB—total volume of tax revenues of the consolidated state budget, monetary units (e.g., rubles);

RCSB—total volume of revenues of the consolidated state budget, monetary units (e.g., rubles);

SETAC + VTL—total volume of expenditures of the consolidated state budget, monetary units (e.g., rubles);

SETAC—total volume of state expenditures for tax administration and control (applied only in case of accessibility of statistical data), monetary units (e.g., rubles);

VTL—violation of tax laws, i.e., total volume of short-received tax revenues of the consolidated state budget (volume of shadow economy), monetary units (e.g., rubles).

Assigning values to non-financial (qualitative) indicators of effectiveness of state tax policy—indicator of stability, transparency, justice, and correspondence to national interests—within the developed method is performed on the basis of expert evaluations with the help of the following scales (Tables 1, 2, 3 and 4).

For treatment of the results of evaluation of effectiveness of state tax policy, we developed a special scale (Table 5).

Table 1 Scale for assigning values to the indicator of stability of state tax policy

Characteristic (qualitative description) of state tax policy over the recent years	Value that is assigned to the indicator of its stability
Tax policy has been changing and is now in the process of reformation	0.5
Tax policy has been reformed several times and is now in the process of reformation	1
Tax policy has been changed slightly and is now in the process of reformation, but reforms are not approved by the population	1.5
Tax policy has been changed slightly and is now in the process of reformation, but reforms are approved by the population	2
Tax policy has been changed slightly and is approved by the population	2.5
Tax policy has remained unchanged and is approved by the population	3

Source Compiled by the authors

Table 2 Scale for assigning values to the indicator of transparency of state tax policy

Characteristic (qualitative description) of state tax policy over the recent years	Value that is assigned to the indicator of its transparency
A lot of taxes and tax regimes, domination of indirect taxes	0.5
A lot of taxes and tax regimes, domination of direct taxes	1
A lot of taxes, domination of indirect taxes, number of tax regimes is minimal	1.5
A lot of taxes, domination of direct taxes, number of tax regimes is minimal	2
Number of taxes and tax regimes is minimal, domination of indirect taxes	2.5
Number of taxes and tax regimes is minimal, domination of direct taxes	3

Source Compiled by the authors

Table 3 Scale for assigning values to the indicator of justice of state tax policy

Characteristic (qualitative description) of state tax policy over the recent years	Value that is assigned to the indicator of its justice
All taxes are subject to proportional taxation that is not approved by the society	0.5
Most taxes are subject to progressive scales of taxation that are not approved by the society	1
All taxes are subject to progressive scales of taxation that are not approved by the society	1.5
All taxes are subject to proportional taxation that is approved by the society	2
Most taxes are subject to progressive scales of taxation that are approved by the society	2.5
All taxes are subject to progressive scales of taxation that are approved by the society	3

Source Compiled by the authors

The advantages of the developed proprietary method of complex evaluation of effectiveness of state tax policy are as follows:

– integration: simultaneous consideration of quantitative (financial) and qualitative (non-financial) indicators of effectiveness of state tax policy;
– applied direction: values of all indicators are calculated separately, which allows determining the general level of effectiveness of state tax policy, determining pros and cons, and developing measures for increasing the effectiveness of this policy.

Table 4 Scale for assigning values to the indicator of correspondence of state tax policy to national interests

Characteristics (qualitative description) of state tax policy over the recent years	Value that is assigned to the indicator of its correspondence to national interests
Taxes restrain development of entrepreneurship, limit export, and lead to increase of import	0.5
Taxes restrain development of entrepreneurship, lead to increase of import, but stimulate development of export	1
Taxes restrain development of entrepreneurship and export, but allow limiting import	1.5
Taxes limit export and lead to increase of import, but stimulate development of entrepreneurship	2
Taxes limit export, but stimulate development of entrepreneurship and limit import	2.5
Taxes stimulate development of entrepreneurship, limit import and stimulate development of export	3

Source Compiled by the authors

Table 5 Scale for treatment of results of evaluation of effectiveness of state tax policy

Direct average of the values of non-financial indicators	Value of the financial indicator of effectiveness of state tax policy				
	≤ 1	(1;1.5]	(1.5;2]	(2;3)	≥ 3
≤ 1	CLE	VLE	VLE	LE	ME
(1;1.5]	VLE	VLE	LE	ME	AAE
(1.5;2]	VLE	LE	LE	AAE	HE
(2;2.5)	LE	ME	AAE	HE	VHE
≥ 2.5	ME	AAE	HE	VHE	MAE

Source Compiled by the authors

CLE critically low effectiveness; *VLE* very low effectiveness; *LE* low effectiveness; *ME* medium effectiveness; *AAE* above average effectiveness; *HE* high effectiveness; *VHE* very high effectiveness; *MAE* maximum effectiveness

We conducted evaluation of effectiveness of tax policy of modern Russia with the help of the developed method based on the data of late 2017—early 2018. The value of the financial indicator of effectiveness of Russian tax policy is calculated in the following way (International Monetary Fund 2018; Federal State Statistics Service 2018): FIestp (2018) = 28,345.1 * (31,046.7/32,395.7)/(166,754 * 0.3848) = 28,345.1 * 0.96/64, 166.94 = 0.42, being in the interval ≤ 1. As a result of expert evaluation of non-financial indicators of effectiveness of Russia's tax policy, the following results were obtained:

- tax policy has been slightly changed and is in the process of reformation, but the reforms are not approved by the population, so the indicator of stability of state tax policy is assigned with the value 1.5;
- a lot of taxes, domination of indirect taxes, number of tax regimes is minimum—so the indicator of transparency of state tax policy is assigned with the value 1.5;
- all taxes are subject to proportional taxation, which is approved by the society, so the indicator of justice of state tax policy is assigned with the value 2;
- taxes restrain development of entrepreneurship and export, but allow limiting import, so the indicator of correspondence of state tax policy to national interests is assigned with the value 1.5.

Direct average of the value of non-financial indicators of effectiveness of the Russian tax policy constituted 1.62 ((1.5 + 1.5 + 2 + 1.5)/4); it is in the interval (1.5;2]. Based on this, effectiveness of tax policy of modern Russia could be characterized as very low. Its advantages are high justice, acceptable stability, transparency, and stimulation of national interests. Weaknesses of modern Russia's tax policy include low financial effectiveness due to large volume of shadow economy.

4 Conclusions

Thus, the offered hypothesis is proved—it is shown that the modern Russia's tax policy is peculiar for low effectiveness. The main reason for this is insufficiently successful implementation of the most important function of the taxation system—provision of collection of taxes for replenishment of state budgets of all levels of the budget system—due to deficit of the consolidated state budget of the RF and critically large volume of tax evasion (shadow economy). Costs of the tax policy exceed its positive results by more than two times, even without consideration of expenditures for state tax administration and control.

In addition to this, state tax policy that is implemented in modern Russia does not fully conform to the announced principles of stability, transparency, justice, and stimulation of national interests. Based on this, it is possible to state that low effectiveness of state tax policy could be one of the reasons of non-optimality of the modern Russia's taxation system. That's why optimization of this system requires increase of effectiveness of state tax policy.

Acknowledgements The reported study was funded by RFBR according to the research project No. 18-010-00103 A.

References

Chalu, H., & Mzee, H. (2018). Determinants of tax audit effectiveness in Tanzania. *Managerial Auditing Journal, 33*(1), 35–63.

Condie, S. S., Evans, R. W., & Phillips, K. L. (2017). Natural limits of wealth inequality and the effectiveness of tax policy. *Public Finance Review, 2*(1), 18–23.

Crespi, G., Giuliodori, D., Giuliodori, R., & Rodriguez, A. (2016). The effectiveness of tax incentives for R&D+i in developing countries: The case of Argentina. *Research Policy, 45*(10), 2023–2035.

Federal State Statistics Service. (2018). Russia in numbers: Short statistical collection. http://www.gks.ru/wps/wcm/connect/rosstat_main/rosstat/ru/statistics/publications/catalog/doc_1135075100641. Data accessed: 14.06.2018.

Ferré, M., Garcia, J., & Manzano, C. (2018). Tax efficiency, seigniorage and Central Bank conservativeness. *Journal of Macroeconomics, 56,* 218–230.

Gashenko, I. V., Zima, Y. S., Stroiteleva, V. A., & Shiryaeva, N. M. (2018). The mechanism of optimization of the tax administration system with the help of the new information and communication technologies. *Advances in Intelligent Systems and Computing, 622,* 291–297.

International Monetary Fund. (2018). Shadow economies around the world: what did we learn over the last 20 years? https://www.imf.org/~/media/Files/.../WP/.../wp1817.ashx. Data accessed: 14.06.2018.

Jang, S., & Eger, R. J. (2018). The effects of state delinquent tax collection outsourcing on administrative effectiveness, efficiency, and procedural fairness. *American Review of Public Administration, 2*(1), 32–37.

Lewis, B. D. (2018). Local government form in Indonesia: tax, expenditure, and efficiency effects. *Studies in Comparative International Development, 53*(1), 25–46.

Muennig, P. A., Mohit, B., Wu, J., Jia, H., & Rosen, Z. (2016). Cost effectiveness of the earned income tax credit as a health policy investment. *American Journal of Preventive Medicine, 51* (6), 874–881.

Popkova, E. G., Bogoviz, A. V., Lobova, S. V., & Romanova, T. F. (2018a). The essence of the processes of economic growth of socio-economic systems. *Studies in Systems, Decision and Control, 135,* 123–130.

Popkova, E. G., Bogoviz, A. V., Ragulina, Y. V., & Alekseev, A. N. (2018b). Perspective model of activation of economic growth in modern Russia. *Studies in Systems, Decision and Control, 135,* 171–177.

Thomson, R. (2017). The effectiveness of R & D tax credits. *Review of Economics and Statistics, 99*(3), 544–549.

Yanıkkaya, H., & Turan, T. (2018). Tax structure and economic growth: Do differences in income level and government effectiveness matter? *Singapore Economic Review, 2*(1), 1–21.

Part IV
Optimization of Taxation at the Level of Separate Economic Subjects

Management of Taxation at a Modern Company: Tax Optimization Versus Tax Load

Aleksei V. Bogoviz, Svetlana V. Lobova, Julia V. Ragulina, Alexander N. Alekseev and Elena I. Semenova

Abstract *Purpose* This work is aimed at studying the essence of taxation management at a modern company, consideration (by the example of modern Russia) of accessible means of tax optimization, determination of related problems, and development of the optimization algorithm of conduct of taxation management at a modern company. *Methodology* The research is based on the existing methodology of evaluation of effectiveness of tax optimization at a company. *Results* In the course of complex analysis of accessible means of taxation optimization at a modern company it is determined that tax optimization is related to emergence of undesired consequences (problems) for a company, most of which have non-tax (production or marketing) nature. Neglecting these consequences distorts the expected results of tax optimization and reduces effectiveness of taxation management at a modern company. *Recommendation* The proprietary optimization algorithm of taxation management at a modern company allows considering not only advantages from tax optimization but also expenditures for its conduct and related risks (problems) for a company, due to which it ensures receipt of precise and authentic estimate results and guarantees high effectiveness of taxation management at a modern company. Usage of the offered optimization algorithm of

A. V. Bogoviz (✉) · J. V. Ragulina · E. I. Semenova
Federal State Budgetary Scientific Institution "Federal Research Center
of Agrarian Economy and Social Development of Rural Areas—All Russian
Research Institute of Agricultural Economics", Moscow, Russia
e-mail: aleksei.bogoviz@gmail.com

J. V. Ragulina
e-mail: julra@list.ru

E. I. Semenova
e-mail: esemenova@bk.ru

S. V. Lobova
Altai State University, Barnaul, Russia
e-mail: barnaulhome@mail.ru

A. N. Alekseev
Financial University Under the Government of the Russian Federation, Moscow, Russia
e-mail: Alexeev_alexan@mail.ru

© Springer Nature Switzerland AG 2019   101
I. V. Gashenko et al. (eds.), *Optimization of the Taxation System: Preconditions,
Tendencies, and Perspectives*, Studies in Systems, Decision and Control 182,
https://doi.org/10.1007/978-3-030-01514-5_12

taxation management at a modern company allows making the practice of tax optimization widely accessible (due to simple and clear calculations) and ensuring well-balanced management of tax and finance, production, and marketing components of entrepreneurial activities, thus leading to growth of global competitiveness of modern Russian companies.

Keywords Taxation management at a modern company · Tax optimization
Tax load · Modern Russia

JEL Clssification E62 · H20 · K34

1   Introduction

Taxation is one of the most important components of entrepreneurial activities. Modern companies face additional difficulties of taxation. The first difficulty is caused by the influence of globalization on modern companies. In the conditions of globalization in striving for optimization of their business processes, modern companies actively conduct foreign economic activities. It could take various forms, from cooperation with foreign suppliers to transnational organization of business and placement of the company's branches in different countries of the world.

One way or another, most of modern companies function within tax systems of two and more countries. If subtleties of taxation in the Russian economic system are well-known and considered, peculiarities of taxation in other countries, where the company conducts foreign economic activities, could lead to additional tax load. This load could be reduced by tax optimization, but it goes beyond the main activities of the company and requires certain efforts.

The second complexity is related to the influence of the global economic crisis, which has turned into stagnation of the global economic system. Increase of deficits of state budgets around the world due to the crisis led to forced change of conditions of taxation of entrepreneurship. Growth of tax load led to new possibilities for tax optimization, which became an inseparable condition for supporting competitiveness of modern companies.

These complexities actualize the necessity for taxation management at a modern company by tax optimization, which is considered to be highly effective. Our hypothesis is that tax optimization and reduction of tax load could lead to undesired consequences (non-tax problems) of a company.

This work is aimed at studying the essence of taxation management at a modern company, consideration (by the example of modern Russia) of accessible means of tax optimization, determination of possible problems, and development of the optimization algorithm of conduct of taxation management at a modern company.

2   Materials and Method

The theoretical and methodological basis of the research consists of the works of modern scholars and experts on the issues of taxation management at a modern company, including tax load and tax optimization: Angelopoulos et al. (2017), Antsyz and Vysotskaya (2018), Assidi et al. (2016), Bustos-Contell et al. (2017), Duan et al. (2018), Gashenko et al. (2017, 2018), Gazda et al. (2017), Hallerberg and Scartascini (2017), Hines (2017), Lesko (2011), Popkova et al. (2018a, b), Qiao and Li (2016), Su and Zheng (2015), Zhang et al. (2018), and Suglobova and Boboshko (2015).

The performed content analysis of scientific literature showed that taxation management at a modern company is a process of determination and practical implementation of perspectives of reduction of its tax load by tax optimization. A criterion of expedience of tax optimization is value of obtained advantages (volume of saving on tax payments). At that, possible negative consequences (problems) of tax optimization for a company are not taken into account.

3   Results

As a result of complex study of the modern Russian practice of taxation management, we determined the following accessible means of taxation optimization at a modern company and related possible problems (Table 1).

Table 1 Accessible means of taxation optimization at a modern company and related possible problems

Corporate tax	Means of optimization of the tax burden	Possible problems that are related to tax optimization
Income tax	Selection of the optimal regime of taxation	Limitation of possibilities of growth and development of business
	Selection of quick type of amortization of fixed assets, growth of expenditures	Short-term effect and further growth of income tax
Custom duties	Limitations of foreign economic activities	Reduction of competitiveness, limitation of possibilities of growth and development of business
Personnel tax (personal income tax and social deductions)	Conclusion of analogs to labor agreements	Low labor efficiency, reduction of competitiveness
	Outsource	
Property taxes	Property rental instead of purchase	Limited possibilities of selection and management of property, reduction of competitiveness
	Property lease	

Source Compiled by the authors

As is seen from Table 1, most corporate taxes have multiple methods of taxation optimization—but all of them are related to future possible problems for the company. Thus, for example, reduction of volume of payments for income tax is possible by means of selection of the optimal regime of taxation. As a rule, this envisages transition to one of special regimes of taxation. Special regimes of taxation in modern Russia are accessible only for subjects of small and medium entrepreneurship, so striving for preservation of belonging to the special regime of taxation in future could limit possibilities of growth and development of business, hindering the increase of the scale of entrepreneurial activities and transition to the level of large entrepreneurship.

Another method of optimization of income tax is selection of a quick type of amortization of fixed assets and growth of income, which allows reducing the volume of taxable income—tax base of income tax. However, this allows obtaining only short-term positive effect and will lead—in the mid-term or long-term—to quick growth of income tax, when amortization of fixed assets will be finished and entrepreneur will be reluctant to bear excessive expenditures, striving for maximization of his income.

In its turn, reduction of the volume of custom duties is possible by limiting the company's foreign economic activities. In the conditions of globalization and active foreign economic activities of rivals, this may lead to reduction of competitiveness and limitation of possibilities of growth and development of business. Personnel taxes (personal income tax and social deductions) could be reduced by means of conclusion of analogs of labor agreements (e.g., agreements of the legal character) and outsource.

The drawbacks of these methods of tax optimization are low labor efficiency and reduction of company's competitiveness. The company's payments for property taxes could be reduced by property rental instead of purchase and by property lease. However, in this case, the possibilities of selection and management of property will be limited—which may lead to reduction of the company's competitiveness.

As is seen, almost in all cases tax optimization provides the company with short-term financial advantages, but leads to production and marketing drawbacks, which in the long-term lead to reduction of business competitiveness and, accordingly, financial losses—reduction of profit and profitability of the company.

The determined problems, which appear due to tax optimization, can make it inexpedient due to low effectiveness (total expenditures and losses' exceeding the volume of total advantages). For preventing inexpedient tax optimization and maximization of effectiveness of taxation management at a modern company, we develop its optimization algorithm (Fig. 1).

As is seen from Fig. 1, unlike the algorithm that is applied in the modern economic practice, the offered algorithm is conducted in five consecutive stages. The first stage remains unchanged and envisages determining the current tax load on the company—in the absolute (total volume of tax expenditures) and relative (share of tax expenditures from revenues and profit of the company) expression.

The second stage is related to determining perspectives of tax optimization—i.e., reduction of tax load on the company. At this stage, various accessible means of

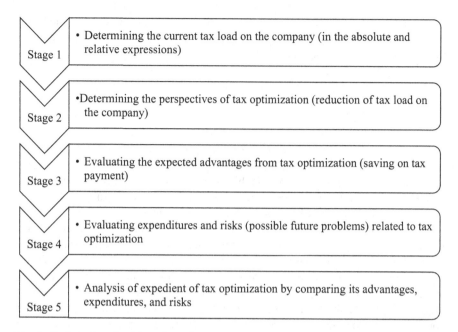

Fig. 1 Optimization algorithm of taxation management at a modern company. *Source* Compiled by the authors

optimization of taxes that are paid by the company are determined and compared. At the third stage, expected advantages from tax optimization are assessed. The company's economy from tax optimization is determined.

This economy is calculated in the absolute expression by finding the difference between current and future tax expenditures of the company. It is calculated in the relative expression—as to revenues—and separately as the share of current tax expenditures. The fourth stage of the developed algorithm is new—it envisages evaluation of expenditures and risks (future possible problems) that are related to tax optimization.

Expenditures for tax optimization include the company's expenditures for document registration of this optimization and expenditures for taxation management, which include wages of tax manager and other expenditures. Risks of tax optimization are to be evaluated by expert method through the prism of their probability and danger for the company. It is recommended to use coefficient of risk of tax optimization, which could exceed or equal 1.

The final, fifth, stage of the developed algorithm of taxation management at a modern company is related to analysis of expedience of tax optimization through comparison of its advantages, expenditures, and risks. During calculations, it is recommended to use the standard formula of effectiveness, in which numerator is economy of tax expenditures (in monetary expression) and nominator is product of expenditures for tax optimization (in monetary expression) and risk coefficient.

Treatment of the received value could be performed in the standard way—the larger the value the more expedient is tax optimization; is the value is below or equals 1, tax optimization is inexpedient.

4 Conclusions

Concluding the performed research, it is possible to note that the offered hypothesis is correct—tax optimization is related to emergence of undesired consequences (problems) for the company, most of which have non-tax (production or marketing) nature. Neglecting these consequences distorts the expected results of tax optimization and reduces effectiveness of taxation management at a modern company.

The developed proprietary optimization algorithm of taxation management at a modern company allows considering advantages from tax optimization and expenditures for its conduct, as well as related risks (problems) for the company, due to which it ensures receipt of precise and authentic estimate results and guarantees high effectiveness of taxation management at a modern company.

Usage of the offered optimization algorithm of taxation management at a modern company will allow making the practice of tax optimization widely accessible (due to simple and clear calculations) and ensuring well-balanced management of tax & financial, production, and marketing components of entrepreneurial activities, thus leading to growth of global competitiveness of modern Russian companies.

References

Angelopoulos, K., Asimakopoulos, S., & Malley, J. (2017). The optimal distribution of the tax burden over the business cycle. *Macroeconomic Dynamics, 2*(1), 1–40.

Antsyz, S. M., & Vysotskaya, T. V. (2018). About some two-level models of optimization of tax schemes. *CEUR Workshop Proceedings, 2098,* 17–32.

Assidi, S., Aliani, K., & Omri, M. A. (2016). Tax optimization and the firm's value: Evidence from the Tunisian context. *Borsa Istanbul Review, 16*(3), 177–184.

Bustos-Contell, E., Climent-Serrano, S., & Labatut-Serer, G. (2017). Offshoring in the European union: A study of the evolution of the tax burden. *Contemporary Economics, 11*(2), 235–248.

Duan, T., Ding, R., Hou, W., & Zhang, J. Z. (2018). The burden of attention: CEO publicity and tax avoidance. *Journal of Business Research, 87,* 90–101.

Gashenko, I., Orobinskaya, I., Orobinskiy, A., Shiryaeva, N., & Zima, Y. (2017). Ways of corporate tax optimisation for cluster entities. *International Journal of Trade and Global Markets, 10*(2–3), 151–159.

Gashenko, I. V., Zima, Y. S., Stroiteleva, V. A., & Shiryaeva, N. M. (2018). The mechanism of optimization of the tax administration system with the help of the new information and communication technologies. *Advances in Intelligent Systems and Computing, 622,* 291–297.

Gazda, J., Kováč, V., Tóth, P., Drotár, P., & Gazda, V. (2017). Tax optimization in an agent-based model of real-time spectrum secondary market. *Telecommunication Systems, 64*(3), 543–558.

Hallerberg, M., & Scartascini, C. (2017). Explaining changes in tax burdens in Latin America: Do politics trump economics? *European Journal of Political Economy, 48,* 162–179.

Hines, J. R. (2017). Business tax burdens and tax reform. *Brookings Papers on Economic Activity, 2017*(Fall), 449–477.

Lesko, M. V. (2011). Adaptation of financial management and tax policies of enterprises to tax code. *Actual Problems of Economics, 1*(7), 260–268.

Popkova, E. G., Bogoviz, A. V., Lobova, S. V., & Romanova, T. F. (2018a). The essence of the processes of economic growth of socio-economic systems. *Studies in Systems, Decision and Control, 135,* 123–130.

Popkova, E. G., Bogoviz, A. V., Ragulina, Y. V., & Alekseev, A. N. (2018b). Perspective model of activation of economic growth in modern Russia. *Studies in Systems, Decision and Control, 135,* 171–177.

Qiao, B., Li, X. (2016). Study on the tax system optimization scheme of the real estate industry based on supply side structural reform. In *ICCREM 2016: BIM Application and Offsite Construction—Proceedings of the 2016 International Conference on Construction and Real Estate Management* (pp. 1019–1027).

Su, L. D., Zheng, J. (2015). A study of the significance of tax planning for the enterprise's financial management. In *Computing, Control, Information and Education Engineering—Proceedings of the 2015 2nd International Conference on Computer, Intelligent and Education Technology, CICET 2015* (pp. 819–822).

Suglobova, A. E., & Boboshko, N. M. (2015). *Taxes and taxation: Study guide* (4th ed., p. 543). Moscow (Russia): UNITI-DANA.

Zhang, C., Cheok, C. K., & Rasiah, R. (2018). The extreme outcomes of corporate tax management: Evidence from Chinese listed enterprises. *Institutions and Economies, 10*(1), 19–52.

Personal Tax Management: Voluntary Initiative or Forced Measure

Alexander E. Suglobov, Oleg G. Karpovich, Elena V. Kletskova, Inna Y. Timofeeva and Tatiana S. Kolmykova

Abstract *Purpose* The purpose of the work is to determine the necessity for individual taxpayers' conducting personal tax management (by the example of modern Russia) and to develop its conceptual model that reflects the logic of this process. *Methodology* For determining the level of necessity for personal tax management, the authors use the method "ad absurdum" with foundation on classic logic. This method is supplemented by a complex of general scientific methods: analysis, synthesis, induction, deduction, and graphic presentation of information (formalization). *Results* The authors analyze ratio of pros and cons of personal tax management and determine that it is necessary in cases of purchase of property, conduct of individual entrepreneurial activities, employment, and decisions on purchase of goods—that is, in most situations faced by a modern individual taxpayer—and allows maximizing the obtained profit, reducing related tax costs, and reducing taxation risks. It is substantiated that personal tax management is a forced measure, which is critically necessary for a modern individual taxpayer. The logic of practical implementation of personal tax management is reflected by the pro-

A. E. Suglobov (✉)
Financial University Under the Government of the Russian Federation,
Moscow, Russia
e-mail: a_suglobov@mail.ru

O. G. Karpovich
Russian Customs Academy, Moscow, Russia
e-mail: iskran@yahoo.com

E. V. Kletskova
Altai State University, Barnaul, Russia
e-mail: stroiteleva_ev@mail.ru

I. Y. Timofeeva
Russian Presidential Academy of National Economy and Public Administration,
Moscow, Smolensk, Russia
e-mail: innatimoff@mail.ru

T. S. Kolmykova
Southwest State University, Kursk, Russia
e-mail: t_kolmykova@mail.ru

© Springer Nature Switzerland AG 2019
I. V. Gashenko et al. (eds.), *Optimization of the Taxation System: Preconditions, Tendencies, and Perspectives*, Studies in Systems, Decision and Control 182,
https://doi.org/10.1007/978-3-030-01514-5_13

109

prietary conceptual model, in which directions and tools of personal tax management are systematized and its purpose and expected results are given. This model is to ensure transparency and clarity of the process of personal tax management for individual taxpayers and make its mass usage possible. *Recommendations* Wide transition from uncontrolled taxation to personal tax management is recommended—it will allow increasing the population's living standards as it will ensure more precise individual budgeting and prevention of individual tax crises related to factual impossibility to pay taxes.

Keywords Personal tax management · Taxation · Individual taxpayer
Modern Russia

JEL Classification E62 · H20 · K34

1 Introduction

In the modern tax systems, individual taxpayers come to the foreground under the influence of two tendencies. The first tendency is that these systems are peculiar for high and increasing complexity. The state strives for simultaneous observation of multiple and crossing functions and principles of taxation and introduces new characteristics of taxes (additional subsidies, etc.) and creates new variations of taxation (additional tax regimes, special categories of taxpayers, etc.), which complicates the formula of calculation of tax load.

The second tendency is that large and increasing share of shadow economy complicates collection of corporate taxes, due to which the state has to increase tax load on population. These tendencies led to the fact that most individual taxpayers do not realize the volume of tax load and cannot forecast the volume of their tax expenditures. Personal tax management has to overcome this uncertainty. However, it is not yet very popular (at least, in modern Russia), due to which it is yet unclear to which extent modern individual taxpayers need it.

We offer a hypothesis that personal tax management is a forced measure that has to be implemented by all individual taxpayers, but that is not widely implemented due to absence of clear understanding of the logic of personal tax management. The purpose of the work is to determine the necessity for individual taxpayers to conduct personal tax management (by the example of modern Russia) and to develop its conceptual modem that reflects the logic of this process.

2 Materials and Method

The performed literature overview on the selected topic showed that personal tax management is a process of individual taxpayers' managing their taxation by determining the current level of tax load, determining and implementing perspectives of its reduction by means of tax optimization, and executing tax obligations.

At that, the necessity for personal tax management is treated ambiguously. Some authors think that its implementation is a voluntary initiative of individual taxpayers, and there is no real necessity in its mass application, for personal tax management is effective only in certain cases. This point of view is presented in the works: Boiko and Drahan (2016), Borba and Coelho (2016), Clingingsmith and Shane (2016), Mohamad et al. (2017), Pratama (2017), Tepperová and Pavel (2016), Williams and Krasniqi (2017), Yilmazkuday (2017), Young et al. (2016), Yusoff and Mohd (2017), Suglobova and Boboshko (2015).

Other scholars think that personal tax management is a forced and highly-effective measure, caused by objective necessity and mandatory for individual taxpayers. This point of view is described in the works: Gashenko et al. (2018), He (2017), Kapoutsou et al. (2015), Nemirova and Tyurina (2015), Popkova et al. (2018a, b), Sawitri et al. (2017), Sundvik (2016), Ying and Geng (2012).

For verification of the offered hypothesis and determining the level of necessity for personal tax management, this research uses the method "ad absurdum" with foundation on the classic logic. This method is supplemented by the complex of general scientific method—analysis, synthesis, induction, deduction, and method of graphic presentation of information.

3 Results

As a result of study of the current Russian laws and applied practice of individual taxation, we determined pros and cons of personal tax management, which ratio is shown in Fig. 1.

Let us view arguments in Fig. 1 in detail. Cons of personal tax management are as follows: taxpayers of indirect taxes are sellers, property taxes are accrued by the tax service, and taxpayers of personal income tax (in case of wages) and social deductions are employers. In view of these arguments, it may seem that possibilities in the sphere of personal tax management are very limited (almost brought down to zero) and there's no necessity for it.

Pros of personal tax management are as follows:

– taxpayer claims his right for tax subsidies himself, and awareness of peculiarities of taxation allows him conducting the necessary economic operations without paying taxes. An example is property buy-sell agreements, which

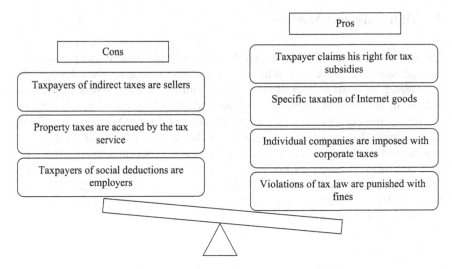

Fig. 1 Ratio of pros and cons of personal tax management. *Source* Compiled by the authors

    envisage that during five years of possession of the property the seller has to pay 13% individual income tax from the sum of the deal, and incomes of the seller are not imposed with the tax during five years;

- specific taxation of Internet goods: large share of E-trade in modern Russia belongs to shadow economy. That's why goods on the Internet are usually cheaper than in stores, as they are imposed with less taxes, and imported goods are even cheaper, as they are not imposed with Russian taxes. However, purchasing goods on the Internet, consumers face increased risk level, as their rights are not fully protected by the state;

- individual companies are imposed with corporate taxes: doing business in the organizational and legal form of individual company envisages emergence of liability for payment of corporate taxes with preservation of the status of individual taxpayer. In this case, responsibility and risks of taxation grow;

- violation of tax law is punished with fines: in all cases, save wages, taxpayers have to file information of incomes to tax authorities with the help of tax declaration and pay individual income tax—otherwise, it will be considered tax evasion. In case of shadow employment, employees are also responsible to receipt of shadow wages, as they do not pay taxes from them.

    In view of the above arguments, it is possible to conclude that without personal tax management individual taxpayer risks a lot—he might violate tax laws and his tax expenditures might exceed total revenues, which, according to the law, is not a reason for tax evasion. The logic of personal tax management is reflected by the developed conceptual model (Fig. 2).

    As is seen from Fig. 2, personal tax management is conducted in four directions for reduction of personal tax risks and tax expenditures. 1st direction: personal

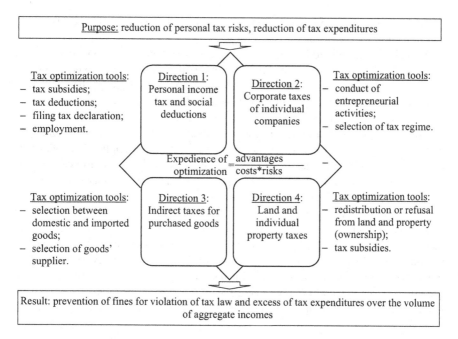

Fig. 2 Conceptual model of personal tax management. *Source* Compiled by the authors

income tax and social deductions. The tools of tax optimization here are tax subsidies, tax deductions, filing tax declaration, and employment (official).

2nd direction: corporate taxes of individual companies. The tools of tax optimization here are conduct of entrepreneurial activities (or refusal from them) and selection of tax regime. 3rd direction: indirect taxes for purchased goods. The tools of tax optimization here are choice between domestic and imported goods and choice of the type of goods' supplier (online store or usual store). 4th direction: land and individual property tax. The tools of tax optimization here are redistribution or refusal from land and property and tax subsidies.

In all directions, the logical formula of decision on conduct of tax optimization is a ratio of expected results from it (economy of tax expenditures) to product of costs (time, efforts, and money) and risks of taxation (fines for tax evasion). As a result, personal tax management is ensures prevention of fines for violation of tax law and excess of tax expenditures over the volume of aggregate incomes.

4 Conclusions

Thus, personal tax management is necessary in cases of purchasing property, conduct of individual entrepreneurial activities, employment, and making decisions on purchase of goods—i.e., in most situations that are faced by a modern individual taxpayer—and allows maximizing the obtained profit, reducing related tax costs, and reducing risks of taxation.

Therefore, the offered hypothesis is correct—personal tax management is a forced measure, which is necessary for a modern individual taxpayer. Logic of practical implementation of personal tax management is reflected by the developed and presented conceptual model, which systematizes directions and tools of personal tax management and shows its goal and expected results.

This model is to ensure transparency and clarity of the process of personal tax management for individual taxpayers and enable its mass application, which is necessary in modern Russia. Transition from uncontrolled taxation to personal tax management will allow increasing the population's living standards, as it will ensure individual budgeting and prevention of individual tax crises, related to factual impossibility to pay taxes.

A precondition to mass application of tax management is the tendency of increase of financial (including tax) awareness of population in modern Russia. It should be noted that personal tax management is aimed not at tax evasion but at its prevention and search for optimal means of conduct of economic activities according to the existing tax laws. That's why its mass application corresponds to interests of the state, as it ensures increase of tax discipline and overcoming of shadow economy.

References

Boiko, S., & Drahan, O. (2016). Individual income tax in the formation of financial resources of the enlarged government. *Economic Annals-XXI, 161*(9–10), 35–38.

Borba, B. E., & Coelho, A. F. C. (2016). Individual income tax, equality and economic capacity: Analyzing the attribute of generality|[Imposto de renda da pessoa física, isonomia e capacidade econômica: Analisando o atributo da generalidade]. *Revista de Investigacoes Constitucionais, 3*(2), 199–223.

Clingingsmith, D., & Shane, S. (2016). How individual income tax policy affects entrepreneurship. *Fordham Law Review, 84*(6), 2495–2516.

Gashenko, I. V., Zima, Y. S., Stroiteleva, V. A., & Shiryaeva, N. M. (2018). The mechanism of optimization of the tax administration system with the help of the new information and communication technologies. *Advances in Intelligent Systems and Computing, 622*, 291–297.

He, J. (2017). Effect of tax burden on income management. In *4th International Conference on Industrial Economics System and Industrial Security Engineering, IEIS 2017*, 8078567.

Kapoutsou, E., Tzovas, C., Chalevas, C. (2015). Earnings management and income tax evidence from Greece. *Corporate Ownership and Control, 12*(2 CONT6), 523–541.

Mohamad, A., Radzuan, N., & Hamid, Z. (2017). Tax arrears amongst individual income taxpayers in Malaysia. *Journal of Financial Crime, 24*(1), 17–34.

Nemirova, G. I., & Tyurina, Y. G. (2015). Ways of possible use of foreign experience in mechanism of tax reduction for individuals in Russia. *Applied Econometrics and International Development, 15*(2), 71–80.

Popkova, E. G., Bogoviz, A. V., Lobova, S. V., & Romanova, T. F. (2018a). The essence of the processes of economic growth of socio-economic systems. *Studies in Systems, Decision and Control, 135*, 123–130.

Popkova, E. G., Bogoviz, A. V., Ragulina, Y. V., & Alekseev, A. N. (2018b). Perspective model of activation of economic growth in modern Russia. *Studies in Systems, Decision and Control, 135,* 171–177.

Pratama, A. (2017). Machiavellianism, perception on tax administration, religiosity and love of money towards tax compliance: Exploratory survey on individual taxpayers in Bandung City, Indonesia. *International Journal of Economics and Business Research, 14*(3–4), 356–370.

Sawitri, D., Perdana, S., Muawanah, U., & Setia, K. A. (2017). The influence of tax knowledge and quality of service tax authorities to the individual taxpayer compliance through taxpayer awareness. *International Journal of Economic Research, 14*(13), 217–235.

Suglobova, A. E., & Boboshko, N. M. (2015). *Taxes and taxation: Study guide* (4th ed., p. 543). Moscow: UNITI-DANA.

Sundvik, D. (2016). Earnings management around Swedish corporate income tax reforms. *International Journal of Accounting, Auditing and Performance Evaluation, 12*(3), 261–286.

Tepperová, J., & Pavel, J. (2016). Evaluation of the impacts of selected tax reforms influencing the income of individuals in the Czech Republic. *Acta Universitatis Agriculturae et Silviculturae Mendelianae Brunensis, 64*(4), 1401–1407.

Williams, C. C., & Krasniqi, B. (2017). Evaluating the individual- and country-level variations in tax morale: Evidence from 35 Eurasian countries. *Journal of Economic Studies, 44*(5), 816–832.

Yilmazkuday, H. (2017). Individual tax rates and regional tax revenues: a cross-state analysis. *Regional Studies, 51*(5), 701–711.

Ying, S., Geng, X. (2012). The management innovation of personal income tax system by comparison between China and Canada. In *Proceeding of 2012 International Conference on Information Management, Innovation Management and Industrial Engineering, ICIII 2012* (vol. 2, pp. 270–273).

Young, A., Lei, L., Wong, B., & Kwok, B. (2016). Individual tax compliance in China: A review. *International Journal of Law and Management, 58*(5), 562–574.

Yusoff, S. N., Mohd, S. (2017). Individual tax compliance decision. *Pertanika Journal of Social Sciences and Humanities, 25*(S), 97–108.

Tax Awareness and "Free Rider" Problem in Taxes

Aleksei V. Bogoviz, Inna N. Rycova, Elena V. Kletskova, Tatyana I. Rudakova and Marina V. Karp

Abstract *Purpose* The purpose of the work is to study the "free rider problem" in taxes and determine the probability of the fact that shadow economy in modern Russia is caused by this problem. *Methodology* For determining the probability of the fact that shadow economy in modern Russia is caused by the "free rider problem" in taxes, the authors conduct a complex logical analysis of information and analytical materials of the Global Financial Literacy Excellence Center, National Research University "Higher School of Economics", the World Bank Group, PricewaterhouseCoopers, Analytical Center "National Agency for Financial Studies", and the All-Russian Public Opinion Research Center, which contain expert evaluations of the level of financial awareness and, in particular, tax awareness in modern Russia, as well as their averaging for obtaining the most realistic picture (as of the 2018 data). *Results* The authors show that the "free rider problem" in taxes is an alternative (opposite) phenomenon to tax opportunism, related to unintentional or insufficiently conscious violation of tax law, which is not

A. V. Bogoviz (✉)
Federal State Budgetary Scientific Institution "Federal Research Center of Agrarian Economy and Social Development of Rural Areas—All Russian Research Institute of Agricultural Economics", Moscow, Russia
e-mail: aleksei.bogoviz@gmail.com

I. N. Rycova
Research Institute of Finance of the Ministry of Finance of the Russian Federation Moscow, Moscow, Russia
e-mail: rycova@yandex.ru

E. V. Kletskova
Altai State University, Barnaul, Russia
e-mail: stroiteleva_ev@mail.ru

T. I. Rudakova
State Budgetary Educational Institution of Higher Education of the Moscow Region "Technological University", Moscow, Russia
e-mail: makmak257@icloud.com

M. V. Karp
State University of Management, Moscow, Russia
e-mail: marvik-09@mail.ru

© Springer Nature Switzerland AG 2019
I. V. Gashenko et al. (eds.), *Optimization of the Taxation System: Preconditions, Tendencies, and Perspectives*, Studies in Systems, Decision and Control 182, https://doi.org/10.1007/978-3-030-01514-5_14

profitable for the state and for tax "free riders". In modern Russia, the level of
financial awareness of the population and, in particular, tax awareness is at the level
of 50%. This means that there's high probability (0.5) that shadow economy in
Russia could be caused by the "free rider problem" in taxes. *Recommendations* For
solving the "free rider problem" in taxes, the authors developed a managerial
concept that ensures reduction of the volume of shadow economy, growth of tax
revenues, overcoming of deficit of state budget, and more successful execution of
state's liabilities (guarantees) before the society.

Keywords Tax awareness · Tax culture · "Free rider problem" in taxes
Modern Russia

JEL Classification E62 · H20 · K34

1 Introduction

The most vivid and generally acknowledged feature of non-optimality of a tax
system is shadow economy—the smaller its scale, the more optimal is the tax
system. The reason for emergence of shadow economy is—according to the economic science—tax opportunism of economic subjects—their conscious opposition
against implemented state tax policy in desire to maximize their own profit (by
means of minimization of tax expenditures).

The concept of shadow economy, based on tax opportunism, is based on the idea
of absolute rationality of economic subjects. However, the modern economic
practice proves that economic subjects are far from absolute rationality—they use
general principles of rationality, but do not possess sufficient possibilities for full
and complete study and analysis of all accessible variants of behavior and selection
of the most optimal one—that is, they are relatively rational.

According to the idea of relative rationality of economic subjects, we offer a
hypothesis that one of the reasons (possible, the most substantial) emergence of
shadow economy, together with tax opportunism, is the "free rider problem" in
taxes, which is rooted in insufficient tax awareness (lack of knowledge of tax law).
The purpose of the work is to study the "free rider problem" in taxes, to determine
the probability of the fact that shadow economy in modern Russia is caused by this
problem, and to develop recommendations for solving it.

2 Materials and Method

The performed overview of modern scientific literature on the topic of shadow
economy showed that in the existing studies and publications the main attention is
paid to determining the essence of shadow economy, development of methodology

of measuring it, and development of measures for its reduction with emphasis on increase of tax administration and control. These include Berger et al. (2018), Blanton et al. (2018), Bogoviz et al. (2018), Buehn et al. (2018), Gashenko et al. (2018), Solis-Garcia and Xie (2018), Veselovsky et al. (2018) and Suglobova and Boboshko (2015).

At that, the only reason for emergence of shadow economy is tax opportunism, and there's no notion of tax "free rider". The topic of tax awareness is studied in the works: Campbell and Helleloid (2016), Floridi (2017), Popkova et al. (2017a, b), and Russell and Brock (2016).

However, it is also connected to the problem of tax opportunism. We think that one-sided and narrow study of shadow economy does not allow for precise determination of the causes of its emergence and for achievement of its full-scale reduction by increase of tax awareness. That's why an alternative vision of the essence of the phenomenon of shadow economy is offered through the prism of "free rider problem" in taxes.

According to our definition, tax "free rider" is a violator of tax discipline, which behavior contradicts the existing tax law—but he cannot fully realize his responsibility for that. Comparing him to transport free rider (public transport passenger who has not paid for ticker) from the positions of scientific psychology, it is possible to distinguish the following variants of behavior of tax "free rider":

- "adventurism": high tax risk appetite, which makes taxpayer check his luck, by not paying taxes;
- "confidence in impunity": confidence that probability of proving the violation of tax law is very low, and the punishment could be avoided;
- "self-justification": preliminary preparation for the case of proving the violation of tax law by looking for a solid justification that allows avoiding the punishment (lack of knowledge of tax law, etc.);
- "lack of initiative": expectation of notification from tax bodies and unawareness of obligations for payment of taxes and showing initiative during taxation (filing tax declaration, etc.).

Tax "free rider" is different from tax opportunist, because tax free rider violates the law with his lack of action, without full awareness of the consequences, while tax opportunist performs a lot of conscious actions, clearly realizing their consequences.

For determining the probability of the fact that shadow economy in modern Russia is caused by the "free rider problem" in taxes, the authors perform a complex logical analysis of information and analytical materials that contain expert evaluations of the level of financial awareness and, in particular, tax awareness in modern Russia, as well as their averaging for receiving the most realistic picture (as of 2018). Its results are shown in Fig. 1.

The data from Fig. 1 show that the modern Russia's tax system is rather complex (47.50 points, 50th position in the world) (World Bank Group, PricewaterhouseCoopers 2018). The level of financial awareness and, in particular,

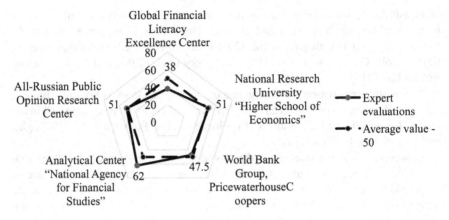

Fig. 1 Expert evaluations and average value of the level of financial awareness and, in particular, tax awareness in modern Russia. *Source* Compiled and calculated by the authors based on: Global Financial Literacy Excellence Center (2018), National Research University "Higher School of Economics" (2018), World Bank Group, Pricewaterhouse Coopers (2018), Analytical Center National Agency for Financial Studies (2018), All-Russian Public Opinion Research Center (2018)

tax awareness, of population constitutes 37% (Global Financial Literacy Excellence Center 2018), 51% (National Research University "Higher School of Economics" 2018; All-Russian Public Opinion Research Center 2018), and 62% (Analytical Center National Agency for Financial Studies 2018).

The calculated direct average is 50%. This means that half of Russians may potentially be tax "free riders". That is, probability of the fact that shadow economy in modern Russia is caused by the "free rider problem" in taxes constitutes 0.5. It should be noted that tax opportunists belong to the second half of the Russian population, as they consciously violate tax law.

3 Results

As a result of studying the "free rider problem" in taxes, we determined that it is a common problem for the state and for taxpayers. Its causal connections in view of these categories of economic subjects are shown in Table 1.

As is seen from Table 1, the causes of emergence of the "free rider problem" in taxes could be imperfection of tax policy—high cost of the tax system, which makes strict observation of tax law difficult, and lack of transparency (including contradiction) of tax policy, which makes it unclear for taxpayers. The causes of emergence of the "free rider problem" in taxes could be irresponsible attitude to tax obligations of taxpayers and low level of financial awareness, in particular, tax awareness.

Table 1 Causal connections of emergence of the "free rider problem" in taxes

	State	Taxpayers
Causes	High complexity of the tax system	Irresponsible attitude to tax obligations
	Lack of transparency of tax policy	Low level of financial awareness and, in particular, tax awareness
Consequences	Shortfall of tax revenues in the state budget	Risk of imposing fines for violation of tax law (in cases of proving the facts of non-payment of taxes)
	High expenditures for tax administration and control	Impossibility of execution of liabilities (guarantees) by the state before taxpayers in full (lack of social protection)

Source Compiled by the authors

The consequences of this problem are negative—at that, for both categories of economic subjects. For the state, they are related to shortfall of tax revenues into the state budgets of all levels of the budget system and, therefore, emergence and increase of budget deficit, as well as high expenditures for tax administration and control, which reduce effectiveness of the tax system.

The consequences for taxpayers, i.e., tax "free riders", consist in risk of fines for violation of tax law (in case of proving the facts of non-payment of taxes) and impossibility for the state to execute its liabilities (guarantees) before taxpayers in full (lack of social protection), which leads to reduction of the population's living standards.

For solving the "free rider problem" in taxes, we developed the following managerial concept (Fig. 2).

As is seen from Fig. 2, the presented managerial concept seeks the goal of preventing the facts of non-payment of taxes and the number of tax "free riders". This goal is achieved by managing the tax behavior of economic subjects. The subject of management is the state in the form of tax service, and the objects of management—economic subjects (taxpayers). Management is conducted in two directions. The first direction: management of tax awareness envisages usage of such tools (implementation of managerial measures) as informing on tax responsibilities and information and consultation support for the issues of tax optimization.

Second direction: management of tax culture—such tools as informing on consequences of violation of tax law—individual (fines) and general (deficit of budget, reduction of living standards, etc.). As a result of implementation of this concept, reduction of the volume of shadow economy, growth of tax revenues, overcoming of the deficit of state budget, and more successful execution of state's obligations (guarantees) before society are achieved.

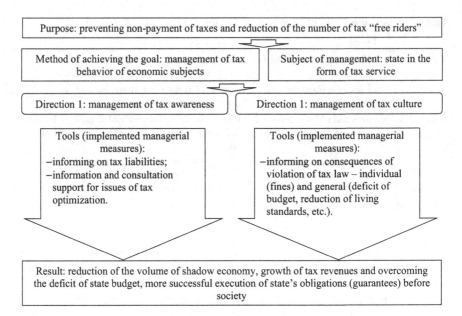

Fig. 2 Managerial concept of solving the "free rider problem" in taxes. *Source* Compiled by the authors

4 Conclusion

Thus, we determined that the "free rider problem" in taxes is an alternative (opposite) phenomenon to tax opportunism, which is related to unconscious violation of tax law that is not profitable for the state and for tax "free riders". In modern Russia, the level of financial awareness of population and, in particular, tax awareness is at the level of 50%. This means that with high probability (0.5) shadow economy in Russia could be caused by the "free rider problem" in taxes.

It should be noted that a restraining factor on the path of increase of financial awareness and, in particular, tax awareness is contradiction of interests of the state. On the one hand, the state is interested in increase of financial awareness and, in particular, tax awareness for provision of taxpayers' awareness of their tax liabilities and, therefore, fuller collection of taxes.

However, on the other hand, with increase of financial awareness and, in particular, tax awareness, taxpayers will understand their tax obligations and tax rights, which will lead to development of civil society and requirements to the state for execution of liabilities before taxpayers. That's why the problem of "free riders" in taxes should be solved not in an isolated way but in complex with other problems within full-scale optimization of taxation in modern Russia.

References

Berger, W., Salotti, S., & Sardà, J. (2018). Do fiscal decentralization and income inequality affect the size of the shadow economy? A panel data analysis for OECD countries. *Applied Economics Letters, 25*(8), 571–575.

Blanton, R. G., Early, B., & Peksen, D. (2018). Out of the shadows or into the dark? Economic openness, IMF programs, and the growth of shadow economies. *Review of International Organizations, 13*(2), 309–333.

Bogoviz, A. V., Ragulina, Y. V., & Sirotkina, N. V. (2018). Systemic contradictions in development of modern Russia's industry in the conditions of establishment of knowledge economy. *Advances in Intelligent Systems and Computing, 622,* 597–602.

Buehn, A., Dell'Anno, R., & Schneider, F. (2018). Exploring the dark side of tax policy: An analysis of the interactions between fiscal illusion and the shadow economy. *Empirical Economics, 54*(4), 1609–1630.

Campbell, K., & Helleloid, D. (2016). Starbucks: Social responsibility and tax avoidance. *Journal of Accounting Education, 37,* 38–60.

Floridi, L. (2017). Robots, jobs, taxes, and responsibilities. *Philosophy and Technology, 30*(1), 28–34.

Gashenko, I. V., Zima, Y. S., Stroiteleva, V. A., & Shiryaeva, N. M. (2018). The mechanism of optimization of the tax administration system with the help of the new information and communication technologies. *Advances in Intelligent Systems and Computing, 622,* 291–297.

Global Financial Literacy Excellence Center. (2018). S&P Global FinLit Survey. http://gflec.org/initiatives/sp-global-finlit-survey/. Data accessed: 16.06.2018.

All-Russian Public Opinion Research Center. (2018). The results of sociological survey of Russians on the issues of financial awareness. https://finoteka.ru/news_page/232. Data accessed: 02.06.2018.

Analytical Center "National Agency for Financial Studies". (2018). Report on financial awareness of Russians—2017. https://nafi.ru/analytics/rossiyane-stali-bolee-tshchatelno-vesti-semeynyy-byudzhet/. Data accessed: 02.06.2018.

National Research University "Higher School of Economics". (2018). Financial awareness of Russians (dynamics and perspectives). https://nnov.hse.ru/data/2012/02/25/1260825213/Kuzina_01_12.pdf. Data accessed: 16.06.2018.

Popkova, E. G., Grechenkova, O. Y., Boris, O. A., Przhedetskaya, N. V., & Gornostaeva, Z. V. (2017a). The prospects of using social marketing in economic criminology in the conditions of an emerging innovative economy. *Russian Journal of Criminology, 11*(2), 280–288.

Popkova, E. G., Lysak, I. V., Titarenko, I. N., Golikov, V., & Mordvintsev, I. A. (2017b). Philosophy of overcoming "institutional traps" and "black holes" within the global crisis management. *Contributions to Economics, 9783319606958,* 321–325.

Russell, H., & Brock, G. (2016). Abusive tax avoidance and responsibilities of tax professionals. *Journal of Human Development and Capabilities, 17*(2), 278–294.

Solis-Garcia, M., & Xie, Y. (2018). Measuring the size of the shadow economy using a dynamic general equilibrium model with trends. *Journal of Macroeconomics, 56,* 258–275.

Suglobova, A. E., & Boboshko, N. M. (2015). *Taxes and taxation: Study guide* (4th ed., p. 543). Moscow (Russia): UNITI-DANA.

Veselovsky, M. Y., Izmailova, M. A., Bogoviz, A. V., Lobova, S. V., & Ragulina, Y. V. (2018). System approach to achieving new quality of corporate governance in the context of innovation development. *Quality—Access to Success, 19*(163), 30–36.

World Bank Group, PricewaterhouseCoopers. (2018). Paying taxes 2018. https://www.pwc.com/gx/en/services/tax/publications/paying-taxes-2018/overall-ranking-and-data-tables.html. Data accessed: 02.06.2018.

Part V
Optimization of Taxation in Modern Russia: The Current State and Tendencies of Reformation

Means of Optimization of the Taxation System with the Help of Informatization and the Problems of Their Application in Russia

Irina V. Gashenko, Yuliya S. Zima and Armenak V. Davidyan

Abstract *Purpose* The purpose of the research is to substantiate the perspectives and to determine the methods and problems of optimization of the taxation system with the help of informatization in modern Russia. *Methodology* For determining the current state of informatization of the modern Russia's economy, the materials of the Global Information Technology Report 2016, provided by the World Economic Forum, are used; for studying dynamics of informatization of Russia's economy—materials of the Measuring the Information Society Reports 2009–2017, provided by the International Telecommunication Union, are used. Regression and correlation analysis is used for determining the influence of informatization of Russia's economy on the index of taxation, calculated by the World Bank. *Results* Results of the performed research showed that modern information and communication technologies allow for full authomatization of the tax system, thus ensuring its large optimization. The developed authors' model is optimized on the basis of new information and communication technologies of the tax system. It is determined that in modern Russia the level of informatization is rather high and continues to grow, stimulating the optimization of the tax system. Most of the methods of informatization of the tax system are applied in modern Russia. The problems related to application of the methods of optimization of the taxation system with the help of informatization in modern Russia include absence of the necessary software and equipment failures. *Recommendations* For further successful optimization of the taxation system with the help of informatization in modern Russia, it is recommended to pay attention to development of infrastructural provision of this process.

I. V. Gashenko (✉) · Y. S. Zima · A. V. Davidyan
Rostov State Economic University (Rostov Institute of National Economy),
Rostov-on-Don, Russian Federation
e-mail: gaforos@rambler.ru

Y. S. Zima
e-mail: zima.julia.sergeevna@gmail.com

A. V. Davidyan
e-mail: dav_121192@mail.ru

© Springer Nature Switzerland AG 2019
I. V. Gashenko et al. (eds.), *Optimization of the Taxation System: Preconditions, Tendencies, and Perspectives*, Studies in Systems, Decision and Control 182, https://doi.org/10.1007/978-3-030-01514-5_15

Keywords Optimization · System of taxation · Informatization
Digital economy · Modern Russia

JEL Classifcation E62 · H20 · K34

1 Introduction

Informatization is one of the new tendencies that dominate in the global economic system. Distribution of modern information and communication technologies—PC, mobile devices, high-speed Internet, etc.—has reached unprecedented scale, due to which not only separate components of economic activities but the whole technological mode change.

This initiates transition of the modern economic systems to a new level of socio-economic development, related to formation of the information (digital) economy. In modern Russia, this process is conducted within the government program "Digital economy", adopted by the Decree dated July 28, 2017, No. 1632-r (Government of the Russian Federation 2018). According to the preliminary result and compiled forecasts, informatization stimulates optimization of economic activities in a range of spheres of national economy.

Based on this, the offered hypothesis is that informatization allows optimizing the system of taxation—but, as the process of formation of the information (digital) economy in modern Russia is at the initial stage (it began in 2017)—informatization of the taxation system faces certain problems. The purpose of the research is to substantiate perspectives and to determine the methods and problems of optimization of the taxation system with the help of informatization in modern Russia.

2 Materials and Method

The conceptual and applied issues of the taxation system informatization are discussed in: Bellalah et al. (2016), Bogoviz et al. (2017), Filipova-Slancheva (2017), Gadžo and Klemenčić (2017), Gashenko et al. (2018), Hamilton and Stekelberg (2017), Heres et al. (2017), Holenstein (2016), Kerzner and Chodikoff (2016), Koroleva and Aleksandrova (2016), Martins (2017), O'Brien (2017), Rohan and Moravec (2017), Solehzoda (2017), and Sukhodolov et al. (2018a, b, c, d).

For determining the current state of informatization of the modern Russia's economy, we use the Global Information Technology Report 2016, provided by the World Economic Forum. For studying dynamics of informatization of Russian economy, we use the materials of Measuring the Information Society Reports 2009–2017, provided by the International Telecommunication Union. Regression and correlation analysis is used for determining the influence of informatization of

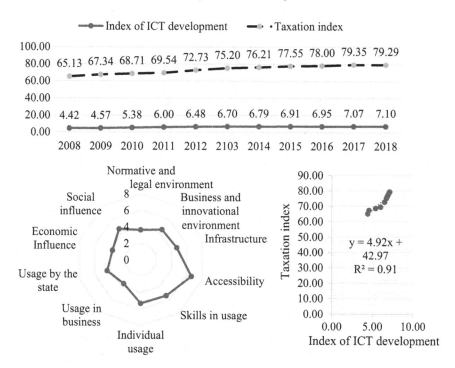

Fig. 1 The current state and dynamics of informatization of the modern Russia's economy and its influence on the tax system. *Source* Compiled by the authors based on: International Telecommunication Union (2018), World Economic Forum (2018), World Bank (2018)

the Russian economy on the taxation index, calculated by the World Bank. The results of the performed analysis are presented in Fig. 1.

As is seen from Fig. 1, the level of informatization of the modern Russia's economy (as of 2018) is rather high. The most prominent parameters of this process are high accessibility of new information and communication technologies and availability of necessary skills. Informatization of the Russia's economy has strong direct (positive) influence on the Russian tax system (correlation—91%). This emphasizes wide perspectives for further optimization of the taxation system of Russia with the help of informatization.

3   Results

Theoretically, the whole tax system could be optimized with the help of new information and communication technologies, which is reflected by the model in Fig. 2.

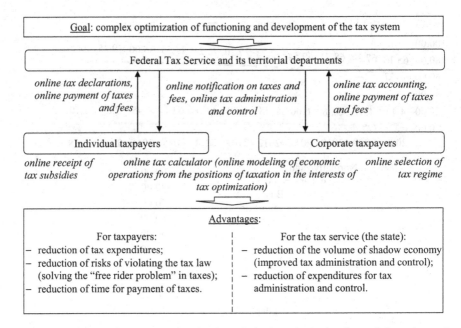

Fig. 2 The model of the tax system that is optimized on the basis of new information and communication technologies. *Source* Compiled by the authors

As is seen from Fig. 2, in the presented model of the tax system that is optimized on the basis of new information and communication technologies, all operations are automatized. The advantages of optimization of the taxation system with the help of informatization include the following. For taxpayers:

– reduction of tax expenditures;
– reduction of risks of violating the tax law (solving the "free rider problem" in taxes);
– reduction of time necessary for payment of taxes.

For the state:

– reduction of the volume of shadow economy (improved tax administration and control);
– reduction of expenditures for tax administration and control.

However, in practice, modern economic systems strive for implementation of the presented model, facing various problems. The methods of optimization of the taxation system with the help of informatization and problems of their application in Russia are shown in Table 1.

As is seen from Table 1, tax operations are subject to informatization from taxpayers and from the state. The methods of optimization of the taxation system with the help of informatization and problems of their application in Russia from taxpayers include the following:

Table 1 Methods of optimization of the taxation system with the help of informatization and problems of their application in Russia

Method of optimization of the taxation system with the help of informatization		Its application in Russia (yes or no)	Problems of application of this method in modern Russia
From taxpayers	Online tax calculator	Yes	Specific features of the tax base for a lot of taxes and impossibility of preliminary calculation
	Online notification of taxpayers on tax liabilities and their execution	Yes	Failures in the work of the tax service's portal
	Online tax declarations	No	Difficulties with confirming the authenticity of documents
	Online receipt of tax subsidies	No	High probability of forgery of documents
	Online tax accounting	No	Absence of software with the tax service
	Online modeling and selection of the tax regime	No	Absence of software
	Online payment of taxes and fees	Yes	Charging additional fee for online payment
From the state	Online determination of facts of tax evasion for personal income tax	Yes	Low level of authomatization—high labor intensity
	Online determination of tax evasion for corporate tax	Yes	Failures in the work of equipment, low tax awareness
	Online inspection of corporate tax accounting (as alternative to field tax inspections)	No	High probability of shadowization of entrepreneurial activities

Source Compiled by the authors

– online tax calculator, which allows planning economic activities in view of its taxation. This method is applied in modern Russia, but with limitations due to specifics of the tax base for a lot of taxes and impossibility for its preliminary determination (e.g., for determining the size of tax on property, it is necessary to know the cadastral number of the taxed property, which is assigned only as a result of corresponding economic operations), which does not allow for full-scale preliminary planning of economic activities for their tax optimization;

– online notification of taxpayers on tax liabilities (sum of tax payments) and their execution (factors of payment or non-payment of taxes). This method is used in modern Russia, but there's a problem of failures in the work of the tax service's portal;

- online tax declaration and online receipt of tax subsidies. These methods are not used in modern Russia, as there's no system of verification of electronic copies of documents;
- online provision of tax accounting. This method is not used in modern Russia (though tax accounting is conducted in the electronic form), as the tax service does not have software for automatic collection and processing of incoming information;
- online modeling and selection of the tax regime in view of specifics of a company. This method is not used in modern Russia due to absence of specialized software.
- online payment of taxes and fees. This method is used in modern Russia, but there's a problem of charging additional fee for online payments.

The methods of optimization of the taxation system with the help of informatization and problems of their application in Russia from the state include the following:

- online determination of facts of tax evasion for personal income tax (by tracking money movement on bank accounts). This method is used in modern Russia, but there's a problem of low level of authomatization of the corresponding operations with the tax service, and, therefore, high labor intensity of the process of tax administration and control;
- online determination of facts of evasion for corporate taxes (e.g., with the help of mandatory cash registers, which inform the tax service on economic operation online). This method is used in modern Russia, but there's a problem of equipment and low tax awareness (not all operations have documentary confirmation);
- online inspection of corporate tax accounting (as an alternative to field tax inspections). This method is not used in modern Russia due to high probability of shadowization of entrepreneurial activities and the necessity for conduct of inventory and other field inspecting measures.

4 Conclusion

The results of the performed research confirmed the offered hypothesis and showed that modern information and communication technologies allow for full authomatization of the tax system, thus ensuring its optimization. In modern Russia, the level of informatization is rather high and continues to grow, thus stimulating optimization of the tax system.

Most of the methods of informatization of the tax system are applied in modern Russia. The problems related to application of the methods of optimization of the taxation system with the help of informatization in modern Russia include absence of the necessary software and failures in work of equipment. Due to this, for further successful optimization of the taxation system with the help of informatization in modern Russia, it is recommended to pay close attention to development of infrastructural provision of this process.

References

Bellalah, M., Bradford, M., & Zhang, D. (2016). A general theory of corporate international investment under incomplete information, short sales and taxes. *Economic Modelling, 58*, 615–626.

Bogoviz, A. V., Ragulina, Y. V., Komarova, A. V., Bolotin, A. V., & Lobova, S. V. (2017). Modernization of the approach to usage of region's budget resources in the conditions of information economy development. *European Research Studies Journal, 20*(3), 570–577.

Filipova-Slancheva, A. (2017). Automatic exchange of tax information: Initiation, implementation and guidelines in Bulgarian context. *Problems and Perspectives in Management, 15*(2–3), 509–516.

Gadžo, S., & Klemenčić, I. (2017). Effective international information exchange as a key element of modern tax systems: Promises and pitfalls of the OECD's common reporting standard. *Public Sector Economics, 41*(2), 207–226.

Gashenko, I. V., Zima, Y. S., Stroiteleva, V. A., & Shiryaeva, N. M. (2018). The mechanism of optimization of the tax administration system with the help of the new information and communication technologies. *Advances in Intelligent Systems and Computing, 622*, 291–297.

Government of the Russian Federation. (2018). Program "Digital economy", adopted by the Decree dated July 28, 2017, No. 1632-r. http://static.government.ru/media/files/9gFM4FHj4PsB79I5v7yLVuPgu4bvR7M0.pdf. Data accessed: 19.06.2018.

Hamilton, R., & Stekelberg, J. (2017). The effect of high-quality information technology on corporate tax avoidance and tax risk. *Journal of Information Systems, 31*(2), 83–106.

Heres, D. R., Kallbekken, S., & Galarraga, I. (2017). The role of budgetary information in the preference for externality-correcting subsidies over taxes: A lab experiment on public support. *Environmental & Resource Economics, 66*(1), 75–82.

Holenstein, D. (2016). Cross-border exchange of information in fiscal offenses: Or the help of the tax office as means of cross-border prosecution of tax offenses [Grenzüberschreitender Informationsaustausch bei Fiskaldelikten: Oder die Steueramtshilfe als Mittel zur grenzüberschreitenden Verfolgung von Steuerdelikten]. *Kriminalistik, 70*(1), 56–62.

International Telecommunication Union. (2018). Measuring the information society reports 2009–2017. https://www.itu.int/en/ITU-D/Statistics/Pages/publications/mis2017.aspx. Data accessed: 19.06.2018.

Kerzner, D. S., & Chodikoff, D. W. (2016). *International tax evasion in the global information age* (pp. 1–425). Springer International Publishing.

Koroleva, L., & Aleksandrova, A. (2016). Information technologies as an instrument to administrate added value tax. *Communications in Computer and Information Science, 674*, 106–122.

Martins, A. F. (2017). Accounting information and its impact in transfer pricing tax compliance: A Portuguese view. *EuroMed Journal of Business, 12*(2), 207–220.

O'Brien, M. (2017). International developments in exchange of tax information. In *Can banks still keep a secret? Bank secrecy in financial centres around the world* (pp. 134–160). Cambridge: Cambridge University Press.

Rohan, J., & Moravec, L. (2017). Tax information exchange influence on Czech based companies' behavior in relation to tax havens. *Acta Universitatis Agriculturae et Silviculturae Mendelianae Brunensis, 65*(2), 721–726.

Solehzoda, A. (2017). Information technologies in the tax administration system of VAT. *Journal of Advanced Research in Law and Economics, 8*(4), 1340–1344.

Sukhodolov, A. P., Popkova, E. G., & Kuzlaeva, I. M. (2018a). Methodological aspects of study of internet economy. *Studies in Computational Intelligence, 714*, 53–61.

Sukhodolov, A. P., Popkova, E. G., & Kuzlaeva, I. M. (2018b). Modern foundations of internet economy. *Studies in Computational Intelligence, 714*, 43–52.

Sukhodolov, A. P., Popkova, E. G., & Kuzlaeva, I. M. (2018c). Peculiarities of formation and development of internet economy in Russia. *Studies in Computational Intelligence, 714*, 63–70.

Sukhodolov, A. P., Popkova, E. G., & Kuzlaeva, I. M. (2018d). Perspectives of internet economy creation. *Studies in Computational Intelligence, 714*, 23–41.

World Bank. (2018). Doing business: Russia. http://russian.doingbusiness.org/data/exploreeconomies/russia. Data accessed: 19.06.2018.

World Economic Forum. (2018). The global information technology report 2016. http://www3.weforum.org/docs/GITR2016/WEF_GITR_Full_Report.pdf. Data accessed: 19.06.2018.

Ways and Methods of Simplification of the Taxation System with the Help of Informatization for Supporting Small and Medium Business

Irina V. Gashenko, Yuliya S. Zima and Armenak V. Davidyan

Abstract *Purpose* The purpose of the work is to determine the influence of informatization of the modern Russian economy on small and medium business and to substantiate the necessity and offer perspective ways and methods of simplification of the Russian taxation system with the help of informatization for supporting small and medium business. *Methodology* For determining the influence of the process of informatization of the modern Russian economy on small and medium business, the authors use regression and correlation analysis. Information and analytical materials of the International Telecommunication Union and the Federal State Statistics Service for 2008–2018 are used. *Results* It is concluded that small and medium business in modern Russia has limited possibilities in the sphere of gaining advantages from informatization of economy and faces the problems—e.g., reduction of competitiveness and growth of expenditures. Tax support for small and medium business in the conditions of informatization is not just possible but necessary. The most perspective ways of implementation of support are authomatization of the tax process, information support for tax optimization, and provision of tax preferences, which are related to informatization, to small and medium business. *Recommendations* The authors recommend institutionalization of simplification of the taxation system with the help of informatization for supporting small and medium business by creating the corresponding normative and legal provision of this process and development and distribution of software.

I. V. Gashenko (✉) · Y. S. Zima · A. V. Davidyan
Rostov State Economic University (Rostov Institute of National Economy),
Rostov-on-Don, Russian Federation
e-mail: gaforos@rambler.ru

Y. S. Zima
e-mail: zima.julia.sergeevna@gmail.com

A. V. Davidyan
e-mail: dav_121192@mail.ru

© Springer Nature Switzerland AG 2019
I. V. Gashenko et al. (eds.), *Optimization of the Taxation System: Preconditions, Tendencies, and Perspectives*, Studies in Systems, Decision and Control 182, https://doi.org/10.1007/978-3-030-01514-5_16

Keywords Simplification of the taxation system · Informatization
Support for small and medium business · Modern Russia

JEL Classification E62 · H20 · K34

1 Introduction

Small and medium business is the basis of successful growth and development of most modern economic systems, including modern Russia. That's why, together with other developed countries, the Russian economy offers preferential conditions for entrepreneurial activities for small and medium business. The most important state tool of creation and support for these conditions is the taxation system.

Two out of four existing (as of 2018) Russian special tax regimes are aimed at small and medium business; the most popular one is "simplified system of taxation". This reflects general logic and essence of tax support for small and medium business, which is related to simplification of the taxation system of this business.

Informatization opens new possibilities for state management of the tax system, and in the context of its general optimization special attention should be paid to the problem of optimization of taxation of small and medium business. At present, there are no special tax preferences for small and medium business, related to usage of new information and communication technologies, in Russia.

This led to offering a hypothesis that the process of informatization of the modern Russia's economy does not perform positive influence on small and medium business, but, on the contrary, can perform negative influence. The purpose of the work is to determine the influence of informatization of the modern Russian economy on small and medium business and to substantiate the necessity and offer perspective ways and methods of simplifying the Russian taxation system with the help of informatization for supporting small and medium business.

2 Materials and Method

Various aspects of support for small and medium business are discussed in the works: Borchers et al. (2016), Bruhn and Loeprick (2016), Fischer-Smith (2018), Freebairn (2017), Gashenko and Zima (2017), Inasius (2018), Ordynskaya et al. (2016), Romli et al. (2017), and Shakirova et al. (2016).

The possibilities that open in the conditions of informatization of modern economic systems, including simplification of the taxation system, are viewed in the publications Gashenko et al. (2018), Krivtsov and Kalimullin (2015), Sukhodolov et al. (2018a, b, c, d, e).

For determining the influence of the process of informatization of the modern Russian economy on small and medium business, the authors use regression and correlation analysis. As the official statistics of the Federal State Statistics Service provide only data for small entrepreneurship, analysis is performed by its example. Dynamics of the values of index of development of the ICT, which reflects the level of economy's informatization, and the number of small companies in Russia in 2008–2018, as well as results of analysis of their dependence, are shown in Fig. 1.

As is seen from Fig. 1, correlation dependence of the number of small companies in Russia in index of ICT development in 2008–2018 is very low (0.03%), which shows the absence of statistically significant connection of these indicators. At that, the determined regression dependence showed negative influence of the process of informatization of the modern Russian economy on development of small business.

This could be caused by the fact that with informatization of economy on the whole and entrepreneurship in particular, small and medium business, which does not have sufficient resources for full-scale modernization, faces the problem of reduction of competitiveness. Striving for overcoming the underrun from large business in the aspect of informatization, small and medium business increases expenditures for its conduct, which leads to bankruptcy.

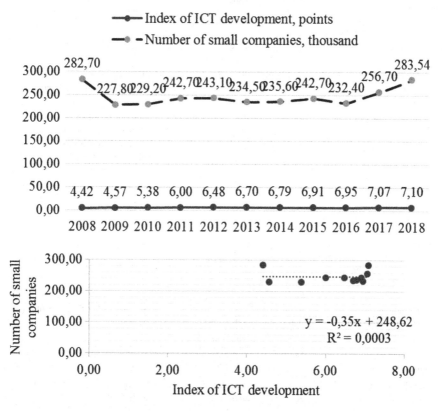

Fig. 1 Dynamics, regression and correlation dependence of the number of small companies in Russia in the index of development of the ICT in 2008–2018. *Source* Compiled by the authors based on: International Telecommunication Union (2018), Federal State Statistics Service (2018)

3 Results

In view of the determined complication of conditions for small and medium business with informatization and high complexity of the tax system of modern Russia, which complicates the process of payment of taxes (differences in terms of payment of different taxes, periodic change of forms of tax documents, etc.), as well as the process of tax optimization (necessity for preliminary selection of the tax regime with the risk of non-optimality as a result of economic activities in the future tax period, etc.), we offer the following perspective (accessible for practical application and potentially popular in modern Russia) ways and methods of simplifying the taxation system with the help of informatization for supporting small and medium business (Table 1).

Table 1 Ways and methods of simplifying the taxation system with the help of informatization for supporting small and medium business

Method	Advantages of the method	Problems of application of the method
Way 1: Automatization of the tax process		
Online document turnover	Electronic storage of tax documents	Necessity for mass informatization of entrepreneurship
Automatic notification of the tax service on business operations	Reduction of labor intensity of tax accounting	Failures in work of equipment (technical devices and Internet connection)
Automatic tax payment	Reduction of the risk of violating the terms of tax payments	Necessity for full-scale automatization of the tax system (inaccessibility with its fragmentary automatization)
Creation and preservation of samples of tax documents	Reduction of labor intensity of tax accounting, saving time spent for payment of taxes	
Filing online tax declaration		
Online receipt and payment for patents		
Way 2: Information support for tax optimization		
Online tax calculator	Reduction of labor intensity and increase of accessibility of tax optimization	Absence of necessary software
Automatic post-fact selection of tax regime	Reduction of tax optimization that is unprofitable for taxpayer	Reduction of tax revenues into the state budget
Way 3: Tax preferences that are related to informatization		
Tax credit for automatization	Modernization of business on the basis of new ICT	Necessity for payment of credit interest
Tax deductions for full authomatization for compensation of expenditures	Automatization of taxation with zero expenditures	Reduction of tax revenues into the state budget

Source Compiled by the authors

As is seen from Table 1, we offer three ways of simplifying the taxation system with the help of informatization for supporting small and medium business. The first way is authomatization of the tax process. It is aimed at simplification of procedures related to taxation for small and medium business and envisages usage of the following methods:

- online document turnover, which envisages electronic registration of all economic operations of the company. It allows eliminating the necessity for storing the supporting tax documents (e.g., on incomes and expenditures for the simplified system of taxation) on paper carriers and transferring to their electronic storage. However, this requires mass informatization of entrepreneurship, which is currently impossible due to technical (low reliability of equipment, etc.) reasons;
- automatic notification of the tax service on business operations. This method is used in modern Russia and allows reducing labor intensity of tax accounting. Its application is complicated by failures in the work of equipment (technical devices and Internet connection);
- automatic tax payment. It envisages automatic payment of taxes and fees (including payment for patent's cost with the patent system of taxation), thus allowing reducing the risk of violation of terms of payment of taxes, creation and preservation of samples of tax documents (receipts, etc.), filing online tax declarations and online receipt and payment for patents, which allows reducing labor intensity of tax accounting and saving time spent for payment of taxes. A restraining factor on the path of application of these methods is the necessity for full-scale automatization of the tax system (inaccessibility with fragmentary automatization).

The second way: information support for tax optimization. It is aimed at simplification and increase of accessibility and effectiveness of optimization of taxation for small and medium business and could be implemented with the help of the following methods:

- online tax calculator, which allows reducing labor intensity and increasing accessibility of tax optimization. Its application requires specialized software;
- automatic post-fact selection of the tax regime. This method envisages that the company conducts electronic document turnover, and its tax accounting is formed automatically. Based on this accounting, after the finish of the tax period, software automatically determines and, which confirmation from the taxpayer, selects the most optimal (profitable from the positions of tax expenditures) regime of taxation. This ensures reduction of risk of tax optimization that is unprofitable for the taxpayer and could lead to reduction of tax revenues into the state budget.

The third way: tax preferences that are related to informatization. It is aimed at simplification and increase of accessibility of the process of informatization of small and medium business by means of taxation and includes the following methods:

- tax credit for automatization. This method envisages delay in payment of tax for purchased new information and communication technologies and equipment (property tax) by the conditions of subsidized crediting (reduces interest for credit as compared to the bank loan). This will allow modernizing small and medium business (which, as a rule, does not have sufficient resources for that) on the basis of new ICT. A drawback of this method is the necessity to pay interest for credit.
- Tax deductions for full authomatization for compensation of expenditures. This method envisages full transition of small and medium business to electronic payments. This will allow achieving maximum transparency of its work, but will require additional expenditures. In order to avoid growth of prices for products of small and medium business due to necessity to pay for online payments, it is expedient to provide tax deductions (for corporate tax). This will ensure authomatization of taxation with zero expenditures and growth of competitiveness of small and medium business; however, this will lead to reduction of tax revenues into the state budget.

It should be noted that these ways and the corresponding methods of simplifying the taxation system with the help of informatization for supporting small and medium business do not contradict each other, but are mutually supplementing. That's why, for achievement of the largest positive effect in modern Russia, related to development of small and medium business, it is recommended to use them systematically.

4   Conclusion

Thus, the offered hypothesis is confirmed; it is substantiated that small and medium business in modern Russia has limited possibilities in the sphere of gaining advantages from informatization of economy and even faces that related problems: reduction of competitiveness and growth of expenditures. Tax support for small and medium business in the conditions of informatization is possible and even necessary.

The most perspective ways of conducting this support include authomatization of the tax process, information support for tax optimization, and providing the small and medium business with tax preferences, related to informatization. The methods of following these ways are accessible for practical implementation in the mid-term (5–10 years).

The necessary telecommunication and human infrastructure is already created in modern Russia. It is recommended to conduct institutionalization of simplification of the taxation system with the help of informatization for supporting small and medium business by creation of the corresponding normative and legal provision of this process, and development and distribution of software.

References

Borchers, E., Deskins, J., & Ross, A. (2016). Can state tax policies be used to grow small and large businesses? *Contemporary Economic Policy, 34*(2), 312–335.

Bruhn, M., & Loeprick, J. (2016). Small business tax policy and informality: Evidence from Georgia. *International Tax and Public Finance, 23*(5), 834–853.

Federal State Statistics Service. (2018). Russia in numbers: Short statistical bulletin. http://www.gks.ru/wps/wcm/connect/rosstat_main/rosstat/ru/statistics/publications/catalog/doc_1135075100641. Data accessed: 20.06.2018.

Fischer-Smith, R. (2018). Adjusting policy implementation frameworks for non-pluralist conditions: A case study of Ukraine's single tax for small business. *Public Administration and Development, 38*(1), 26–38.

Freebairn, J. (2017). Comparison of a lower corporate income tax rate for small and large businesses. *eJournal of Tax Research, 15*(1), 4–21.

Gashenko, I. V., & Zima, Y. S. (2017). Perspectives of development of small business within modernization of tax and cluster policy. In *Contributions to Economics* (pp. 529–536). ISBN 9783319454610.

Gashenko, I. V., Zima, Y. S., Stroiteleva, V. A., & Shiryaeva, N. M. (2018). The mechanism of optimization of the tax administration system with the help of the new information and communication technologies. *Advances in Intelligent Systems and Computing, 622*, 291–297.

Inasius, F. (2018). Factors influencing SME tax compliance: Evidence from Indonesia. *International Journal of Public Administration, 2*(1), 1–13.

International Telecommunication Union. (2018). Measuring the information society reports 2009–2017. https://www.itu.int/en/ITU-D/Statistics/Pages/publications/mis2017.aspx. Data accessed: 20.06.2018.

Krivtsov, A. I., & Kalimullin, D. M. (2015). The model of changes management information system construction. *Review of European Studies, 7*(2), 10–14.

Ordynskaya, M. E., Silina, T. A., Karpenko, S. V., & Divina, L. E. (2016). Tax incentives for small and medium businesses in European union countries in the crisis period. *International Journal of Economics and Financial Issues, 6*(2), 212–218.

Romli, M. A., Nassir, A. M., Kamarudin, F., & Nasir, A. M. (2017). Tax expenditure analysis and reporting: An analysis of selected SME tax programs in Malaysia. *International Journal of Economic Research, 14*(15), 151–175.

Shakirova, R. K., Kurochkina, N. V., & Nikolayeva, L. V. (2016). The meaning and essence of the simplified taxation system in the Russian federation as a tool of small business subjects' tax support. *International Journal of Economic Perspectives, 10*(2), 228–233.

Sukhodolov, A. P., Popkova, E. G., & Kuzlaeva, I. M. (2018a). Methodological aspects of study of internet economy. *Studies in Computational Intelligence, 714*, 53–61.

Sukhodolov, A. P., Popkova, E. G., & Kuzlaeva, I. M. (2018b). Modern foundations of internet economy. *Studies in Computational Intelligence, 714*, 43–52.

Sukhodolov, A. P., Popkova, E. G., & Kuzlaeva, I. M. (2018c). Peculiarities of formation and development of internet economy in Russia. *Studies in Computational Intelligence, 714*, 63–70.

Sukhodolov, A. P., Popkova, E. G., & Kuzlaeva, I. M. (2018d). Perspectives of internet economy creation. *Studies in Computational Intelligence, 714*, 23–41.

Sukhodolov, A. P., Popkova, E. G., & Kuzlaeva, I. M. (2018e). Production and economic relations on the internet: Another level of development of economic science. *Studies in Computational Intelligence, 714*.

Informatization as a Mechanism of Fighting Tax Evasion

Chinara R. Kulueva, Pirmat K. Kupuev
and Mirlanbek B. Ubaidullaev

Abstract *Purpose* The purpose of the research is to study the influence of informatization on shadow economy in Russia and to substantiate the perspectives and to develop recommendations for using informatization as a mechanism of fighting tax evasion. *Methodology* Methodology of the research is based on application of the methods of horizontal, trend, and regression and correlation analysis for studying dynamics and dependence of the volume of shadow economy in Russia on the index of ICT development in 2008–2018. *Results* As a result of the research, it is substantiated that modern Russia has large potential of overcoming the shadow economy by means of the mechanism of tax evasion on the basis of informatization, which is showed by the developed authors' concept of automatized tax administration and control within fighting tax evasion. At present, this potential is not sufficiently opened due to lack of regulation of the process of informatization from the positions of taxation. That's why with informatization of the Russian economic system, the volume of shadow economy increases—instead of expected reduction. This problem could be solved with the given concept—which offers a complex of measures for transfer all companies' payments into the electronic form, thus ensuring their transparency and control. *Recommendations* For successful practical implementation of the developed concept, it is recommended to provide technical possibilities for full-scale authomatization of payments in economy, in order to avoid payment crisis. Also, it is necessary to consider the initial reasons for shadowization of entrepreneurial activities, which are not related only to low accessibility of state services—but also high tax load. Therefore, in order to overcome shadow economy, measures for informatization, which allow eliminating

C. R. Kulueva (✉) · P. K. Kupuev · M. B. Ubaidullaev
Osh State University, Osh, Kyrgyzstan
e-mail: ch.kulueva@mail.ru

P. K. Kupuev
e-mail: ch.kulueva@mail.ru

M. B. Ubaidullaev
e-mail: u.mirlanbek@mail.ru

© Springer Nature Switzerland AG 2019

143

I. V. Gashenko et al. (eds.), *Optimization of the Taxation System: Preconditions, Tendencies, and Perspectives*, Studies in Systems, Decision and Control 182, https://doi.org/10.1007/978-3-030-01514-5_17

the possibility for shadowization of business are not enough—there's a necessity for measures for reduction of tax load on business, which makes tax evasion inexpedient.

Keywords Informatization · Fighting tax evasion · Tax opportunism
Shadow economy · Modern Russia

JEL Classification E62 · H20 · K34

1   Introduction

Informatization is a complex and contradictory process, which may lead to completely opposite consequences. On the one hand, a manifestation of informatization is increase of transparency and controllability of economic processes and systems. Transferring economic operations into the electronic form, informatization allows recording their completion and opens possibilities for automatized monitoring, analysis, and management.

On the other hand, informatization leads to increase of vulnerability of economic processes and systems to negative internal and external influences—such as failures in work of equipment, cyber-attacks, software failures, etc. Due to this, informatization economic processes and systems are susceptible to crises—and thus their control and management is very difficult.

This contradiction of informatization leads to high level of its unpredictability and risk. Spontaneous development of the process of informatization is inadmissible, as it will probably lead to the crisis of the object of informatization. Therefore, it is necessary to regulate the process of informatization for maximization of its effectiveness (reduction of threats and maximization of the obtained advantages).

In modern Russia, informatization of entrepreneurial activities is not fully considered in the state tax policy. Thus, a hypothesis is offered—potential of informatization in the sphere of overcoming shadow economy is insufficiently developed in modern Russia. The purpose of the research is to study the influence of informatization on shadow economy in Russia and to substantiate perspectives and to develop recommendations for using informatization as a mechanism of fighting tax evasion.

2   Materials and Method

As the "free rider" problem in taxes and perspectives of its solution in modern Russia are viewed in previous chapters, this chapter is devoted to the problem of tax opportunism (entrepreneurial structures). This phenomenon is related to conscientious and carefully planned tax evasion as a protest against the conducted state tax policy.

The problem of tax opportunism is studied in detail in the existing works of modern authors: Duan et al. (2018), Janský (2018), Jiménez-Angueira (2018), Lorenz (2018), and Wagner (2018). The authors also use scientific works in the sphere of process of economy's informatization: Gashenko et al. (2018), Krivtsov and Kalimullin (2015), and Sukhodolov et al. (2018a, b, c, d, e).

Methodology of the research includes horizontal, trend, regression, and correlation analysis, which are used for studying dynamics and dependence of the volume of shadow economy in Russia on the index of ICT development in 2008–2018 (Fig. 1).

As is seen from Fig. 1, in 2008–2018 there has been vivid and sustainable growth of the volume of shadow economy in Russia. Trend (growth as compared to 2008) in 2018 constituted 32.61%. Regression analysis showed that increase of the value of the index of ICT development by 1 point leads to growth of the volume of shadow economy in Russia by 6.9939%. Correlation of these indicators is moderate and constitutes 38.16%. Therefore, informatization increases and accelerates the process of shadowization of modern Russia's economy, which is caused by insufficient state regulation of informatization.

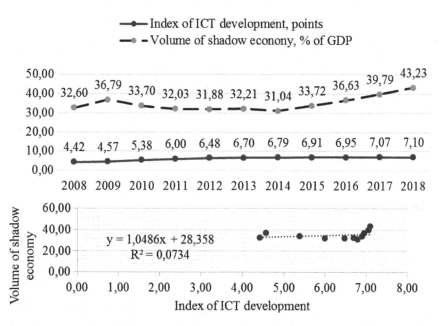

Fig. 1 Dynamics and regression and correlation dependence of the volume of shadow economy in Russia on the index of ICT development in 2008–2018. *Source* Compiled by the authors based on: International Telecommunication Union (2018)

3 Results

As a result of studying the modern Russian practice of conducting entrepreneurial activities, we determined the following causes of tax opportunism of entrepreneurial structures:

- high complexity of doing official business due to inaccessibility of necessary state services (complexity and resource intensity of registration of entrepreneurial activities);
- companies' striving for improvement of financial indicators of their activities (growth of profit and profitability) by means of increase of competitiveness (pricing advantages) as compared to official business;
- unprofitability and commercial unattractiveness of official business (high level of taxes).

These reasons could be eliminated by the developed concept of automatized tax administration and control within fighting tax evasion (Fig. 2).

As is seen from Fig. 2, the developed concept is aimed at provision of successful (highly-effective) struggle against tax evasion and overcoming of shadow economy. The mechanism for achievement of this goal is complex informatization of the aspects of entrepreneurial activities, related to taxation. This mechanism includes

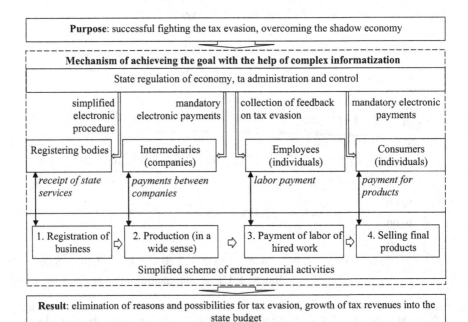

Fig. 2 The concept of automatized tax administration and control within fighting tax evasion. *Source* Compiled by the authors

the following measures of informatization of state regulation of economy, tax administration, and control:

- simplified electronic procedure of registration of business, which allows increasing accessibility of the corresponding state services and fighting tax evasion of shadow (not registered officially) business;
- requirement for mandatory execution of electronic payments between companies and consumers (individuals), which allows fighting evasion from added value tax, excises, and corporate tax;
- electronic collection of feedback from employees on employers' tax evasion, which allows fighting tax evasion for personal income tax (from employees' wages) and social deductions.

As a result, causes and possibilities for tax evasion are eliminates, and growth of tax revenues into the state budget is achieved. The process of overcoming shadow economy by means of fighting tax evasion on the basis of informatization should be performed in three consecutive stages (Fig. 3).

As is seen in Fig. 3, at the first stage (2019–2020) favorable conditions for voluntary de-shadowization of business are created by simplifying the procedure of its registration and initial (test) implementation of measures of automatized tax administration and control within fighting tax evasion. At the second stage (2021–2022), possibilities for shadow business are eliminated, as the tested measures come into effect.

At the third stage (2023–2025), when growth of tax revenues into the budget is achieved, the state reduces taxes for prevention of repeated shadowization of business. According to this, the essence of the process of overcoming the shadow economy is shown in Fig. 4.

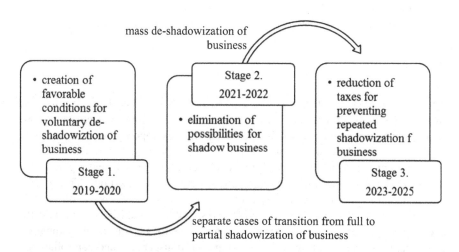

Fig. 3 Stages of the process of overcoming shadow economy with the mechanism of fighting tax evasion on the basis of informatization. *Source* Compiled by the authors

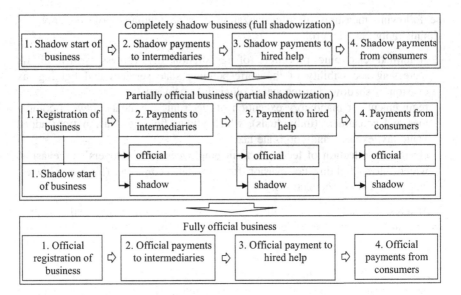

Fig. 4 Essence of the process of overcoming the shadow economy. *Source* Compiled by the authors

As is seen from Fig. 4, fully shadow business, which is not officially registered and which conducts shadow payments with intermediaries, consumers, and workers, transforms into partially official business. Due to simplified electronic registration of the whole or of the part of economic activities, the business conducts shadow and official payments to intermediaries, consumers, and employees. Then, partially official business becomes fully official, ceasing to conduct shadow payments to interested parties.

4   Conclusion

Thus, the offered hypothesis is proved—modern Russia has large potential of overcoming the shadow economy by means of the mechanism of tax evasion on the basis of informatization, which is shown by the developed authors' concept of automatized tax administration and control within fighting tax evasion.

At present, this potential is not fully developed due to insufficient regulation of the process of informatization from the positions of taxation. That's why informatization of the Russian economic system leads to increase of shadow economy—instead of expected reduction. This problem could be solved with the above concept, which offers a complex of measures for transfer of all company's payments into electronic form, thus ensuring their transparency and controllability.

At that, it is necessary to ensure technical means for full-scale authomatization of payments in economy, for avoiding payment crisis. Also, it is necessary to consider initial reasons for shadowization of entrepreneurial activities, related not only to low accessibility of state services but also high tax load. Therefore, for overcoming the shadow economy it is necessary to conduct measures for reduction of tax load on business, which makes tax evasion inexpedient.

It should be concluded that the advantage of the developed authors' concept of automatized tax administration and control within fighting tax evasion is the possibility of its practical implementation and receipt of substantial results in the mid-term (by 2025), which is shown by the described stages of the process of overcoming the shadow economy by means of the mechanism of fighting tax evasion on the basis of informatization.

References

Duan, T., Ding, R., Hou, W., & Zhang, J. Z. (2018). The burden of attention: CEO publicity and tax avoidance. *Journal of Business Research, 87*, 90–101.

Gashenko, I. V., Zima, Y. S., Stroiteleva, V. A., & Shiryaeva, N. M. (2018). The mechanism of optimization of the tax administration system with the help of the new information and communication technologies. *Advances in Intelligent Systems and Computing, 622*, 291–297.

International Telecommunication Union. (2018). Measuring the information society reports 2009–2017. https://www.itu.int/en/ITU-D/Statistics/Pages/publications/mis2017.aspx. Data accessed: 20.06.2018.

Janský, P. (2018). Estimating the revenue losses of international corporate tax avoidance: The case of the Czech Republic. *Post-Communist Economies, 2*(1), 1–19.

Jiménez-Angueira, C. E. (2018). The effect of the interplay between corporate governance and external monitoring regimes on firms' tax avoidance. *Advances in Accounting, 41*, 7–24.

Krivtsov, A. I., & Kalimullin, D. M. (2015). The model of changes management information system construction. *Review of European Studies, 7*(2), 10–14.

Lorenz, J. (2018). Population dynamics of tax avoidance with crowding effects. *Journal of Evolutionary Economics, 2*(1), 1–29.

Sukhodolov, A. P., Popkova, E. G., & Kuzlaeva, I. M. (2018a). Methodological aspects of study of internet economy. *Studies in Computational Intelligence, 714*, 53–61.

Sukhodolov, A. P., Popkova, E. G., & Kuzlaeva, I. M. (2018b). Modern foundations of internet economy. *Studies in Computational Intelligence, 714*, 43–52.

Sukhodolov, A. P., Popkova, E. G., & Kuzlaeva, I. M. (2018c). Peculiarities of formation and development of internet economy in Russia. *Studies in Computational Intelligence, 714*, 63–70.

Sukhodolov, A. P., Popkova, E. G., & Kuzlaeva, I. M. (2018d). Perspectives of internet economy creation. *Studies in Computational Intelligence, 714*, 23–41.

Sukhodolov, A. P., Popkova, E. G., & Kuzlaeva, I. M. (2018e). Production and economic relations on the internet: Another level of development of economic science. *Studies in Computational Intelligence, 714*.

Wagner, F. W. (2018). Tax avoidance and corporate social responsibility [Steuervermeidung und Corporate social responsibility]. *Perspektiven der Wirtschaftspolitik, 19*(1), 2–21.

Part VI
Top-Priority Directions of Optimization of Taxation in Modern Russia

The Model of Development and Implementation of Effective Tax Policy in Modern Russia

Tatyana N. Litvinova

Abstract *Purpose* The purpose of the research is to determine the problems of the algorithm of development and implementation of tax policy that is applied in modern Russia and to develop recommendations for its improvement, as well as to compile a model of development and implementation of effective tax policy in modern Russia. *Methodology* The authors use the method of problem analysis for determining the algorithm of development of implementation of tax policy that is applied in modern Russia; its stages and drawbacks, which are reasons for non-optimality of the tax system, are determined. *Results* The authors determine a problem that is related to incompleteness of criteria of evaluation of effectiveness of state tax policy—absence of criterion of influence on the shadow economy—and insufficiency of evaluation of effectiveness, which does not allow determining the signs of reduction of effectiveness of measures of tax policy in the process of their practical implementation and canceling or terminating their application. An alternative is the developed perspective algorithm of development of implementation of effective tax policy, which envisages application of the criterion of influence on the shadow economy during evaluation of effectiveness of tax policy measures and double evaluation of effectiveness of these measures—preliminary and in the process of their practical implementation. *Recommendations* Based on the determined reason for non-optimality of the modern Russia's tax system, related to shadow economy, the authors' model of development of implementation of effective tax policy in modern Russia is developed—it is oriented at the mid-term and allows eliminating this reason, stimulating the optimization of the Russian tax system.

Keywords Development and implementation of effective tax policy Optimization · Modern Russia

JEL Classification E62 · H20 · K34

T. N. Litvinova (✉)
Volgograd State Agrarian University, Volgograd, Russian Federation
e-mail: litvinova1358@yandex.ru

© Springer Nature Switzerland AG 2019
I. V. Gashenko et al. (eds.), *Optimization of the Taxation System: Preconditions, Tendencies, and Perspectives*, Studies in Systems, Decision and Control 182,
https://doi.org/10.1007/978-3-030-01514-5_18

1   Introduction

High effectiveness is the most important condition of expedient of making decisions and implementation of state tax policy. The signs of non-optimality of modern Russia's tax system, determined in previous chapters, show the necessity for increasing the effectiveness of state tax policy, which is stimulates by controlled informatization of the Russian economy on the whole and the tax system in particular.

It should be noted that while conducting tax policy the state is always guided by desire to achieve its highest possible effectiveness. However, despite commonness of fundamental settings, there are a lot of methodological variations of treatment of effectiveness of state tax policy. Selection of the optimal treatment for each separate socio-economic system should be conducted in view of peculiarities of its functioning at present and goals of its development in future.

In previous chapters, we developed the most optimal treatment of effectiveness of state tax policy, adapted to specifics of modern Russia and purposes of its socio-economic development in the mid-term. That's why from the positions of the offered treatment, it is expedient to evaluate the level of effectiveness of the Russian tax system and the developed and implemented state tax policy.

The working hypothesis is that the reason of non-optimality of the modern Russia's tax system is incomplete consideration of the offered criteria of effectiveness during development and implementation of state tax policy. The purpose of the research is to determine the problems of modern Russia's algorithm of development of implementation of tax policy and to develop recommendations for its improvement, as well as to compile a model of development of implementation of effective tax policy in modern Russia.

2   Materials and Method

The theoretical basis of the research includes theoretical and applied studies of modern scholars, devoted to the issues of evaluation and achievement of high effectiveness during development and implementation of tax policy of the state (Chan et al. 2017; Ferré et al. 2018; Frondel and Vance 2018; Huang et al. 2017; Nwokenkwo 2016; Rubio et al. 2017; Ryu and Keum 2016; Waluyo 2018) and specifics of implementation of this process in modern Russia (Gashenko et al. 2018; Kovalev et al. 2016; Krivtsov and Kalimullin 2015; Musaeva et al. 2015a, b; Popkova et al. 2018a, b).

As a result of application of the method of problem analysis, we determined the algorithm of development of implementation of tax policy, which is applied in modern Russia, which includes the following consecutive stages:

- determining actual drawbacks and needs of the tax system;
- developing perspective measures for overcoming drawbacks and satisfying the needs of the tax system;
- preliminary evaluation of effectiveness of the developed measures with application of such criteria of financial effectiveness as the volume of tax payments (revenues of state budgets of all levels) and expenditures for tax administration and control;
- preliminary evaluation of effectiveness of the developed measures with application of such criteria of non-financial effectiveness as stability, transparency, justice, and correspondence to national interests.

As is seen, according to the applied algorithm in the process of development of implementation of effective state tax policy, shadow economy is not taken into account during evaluation of financial effectiveness. This is the most probable reason of high and increasing volume of shadow economy in Russia. Another problem of the applied algorithm is momentaneousness of evaluation of effectiveness of state tax policy, as in the process of its practical implementation there could appear unexpected signs of ineffectiveness, but its cancelling after preliminary approval is not envisaged by this algorithm.

3 Results

The determined problems could be overcome by the developed algorithm of development of implementation of effective tax policy (Fig. 1).

As is seen from Fig. 1, in the offered perspective algorithm of development of implementation of effective tax policy, shadow economy is one of the criteria of financial effectiveness of developed and implemented measures of this policy. Also, double evaluation of effectiveness of state tax policy is envisaged—preliminary (at the stage of development) and in the process of implementation.

It is recommended to use the possibilities of new information and communication technologies for mass involvement of interested parties in the process of discussion of state tax policy. It is offered to allow everybody to evaluate effectiveness of planned and implemented measures of state tax policy in a private cabinet at the portal of state services.

This will ensure feedback collection and full harmonization of interests of the state and taxpayers in developed and implemented tax policy. At present, implementation of the following model of development of implementation of effective tax policy in modern Russia is expedient (Fig. 2).

As is seen from Fig. 2, the developed model offers complex tools aimed at increase of effectiveness of state tax policy in modern Russia, which envisages:

- increase of tax administration and control (through automatization), by introducing requirements to mandatory character of online payments and

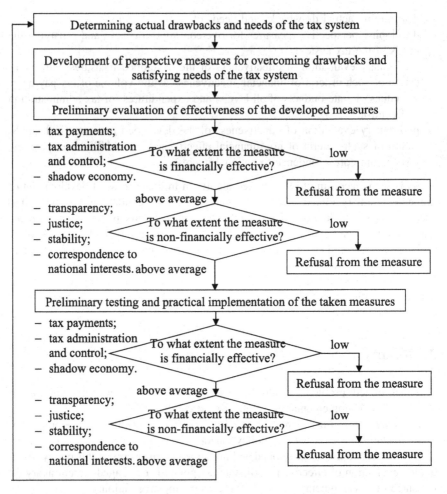

Fig. 1 A perspective algorithm of development of implementation of effective tax policy. *Source* Compiled by the authors

modernization of technologies and equipment of tax administration and control on the basis of new information and communication technologies;
- increase of punishments for tax evasion by increase of fines for violation of tax laws;
- stimulation of increase of tax awareness and tax responsibility of population, by wide information and consultation support;
- simplification of official business and stimulation of barrier-free (stimulated) de-shadowization by provision of the possibility for electronic registration of business and electronic taxation;
- expansion of the specter and simplification of receipt of state guarantees—e.g., unemployment benefits, pensions, social payments, etc.

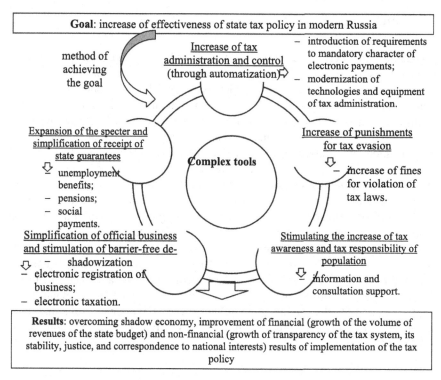

Fig. 2 The model of development of implementation of effective tax policy in modern Russia. *Source* Compiled by the authors

The presented model takes into account the current state of the modern Russia's tax system and is oriented at overcoming its main problem—shadow economy. The offered model of development of implementation of effective tax policy in modern Russia ensures inexpedience of tax evasion, as the value of advantages from observation of tax laws (elimination of the risk of application of punishments for its violation, provided state guarantees, and public reputation) exceeds the tax load.

As a result, overcoming of shadow economy and improvement of financial (growth of revenues of the state budget) and non-financial (growth of transparency of the tax system, its stability, justice, and correspondence to national interests) results of implementation of tax policy and optimization of tax system of modern Russia are ensured.

4   Conclusion

Concluding the performed research, it is possible to state that the working hypothesis is proved—modern Russia does apply the algorithm of development of implementation of state tax policy, which does not fully correspond to the determined criteria of effectiveness of this policy. In particular, there's a problem related to incompleteness of criteria of evaluation of effectiveness of state tax policy—absence of the criterion of influence on shadow economy—and momentaneousness of evaluation of effectiveness, which does not allow determining the signs of reduction of effectiveness of the measures of tax policy in the process of their practical implementation or cancelling or terminating them.

An alternative is the developed perspective algorithm of development of implementation of effective tax policy, which envisages application of the criterion of influence on shadow economy during evaluation of effectiveness of measures of tax policy and double evaluation of effectiveness of these measures—preliminary and in the process of their practical implementation. Based on the determined reason for non-optimality of the tax system of modern Russia, related to shadow economy, we developed and presented a proprietary model of development of implementation of effective tax policy in modern Russia, oriented at the mid-term, which allows eliminating this reason, thus stimulating the optimization of the Russian tax system.

References

Chan, S.-G., Ramly, Z., & Karim, M. Z. A. (2017). Government spending efficiency on economic growth: Roles of value-added tax. *Global Economic Review, 46*(2), 162–188.

Ferré, M., Garcia, J., & Manzano, C. (2018). Tax efficiency, seigniorage and central bank conservativeness. *Journal of Macroeconomics, 56,* 218–230.

Frondel, M., & Vance, C. (2018). Drivers' response to fuel taxes and efficiency standards: Evidence from Germany. *Transportation, 45*(3), 989–1001.

Gashenko, I. V., Zima, Y. S., Stroiteleva, V. A., & Shiryaeva, N. M. (2018). The mechanism of optimization of the tax administration system with the help of the new information and communication technologies. *Advances in Intelligent Systems and Computing, 622,* 291–297.

Huang, S.-H., Yu, M.-M., Hwang, M.-S., Wei, Y.-S., & Chen, M.-H. (2017). Efficiency of tax collection and tax management in Taiwan's local tax offices. *Pacific Economic Review, 22*(4), 620–648.

Kovalev, A. S., Koltsova, T. A., Pelkova, S. V., Khairullina, N. G., & Malezkij, A. A. (2016). Efficiency evaluation of on-site tax audits in Russia: Regional aspect. *Man in India, 96*(10), 3969–3980.

Krivtsov, A. I., & Kalimullin, D. M. (2015). The model of changes management information system construction. *Review of European Studies, 7*(2), 10–14.

Musaeva, K. M., Aliev, B. K., Alieva, E. B., Magomedtagirov, M. M., & Imanshapieva, M. M. (2015a). Problems of implementation of principles of social justice and economic efficiency in the mechanism of charging personal income tax in the Russian Federation. *International Journal of Economics and Financial Issues, 5*(3S), 105–112.

Musaeva, K. M., Suleimanov, M. M., Isaeva, S. M., & Pinskaya, M. R. (2015b). Improvement of the efficiency of the instruments of tax regulation in the context of the development of fiscal federalism and strengthening of the taxable capacity of the subjects of the Russian Federation. *Ecology, Environment and Conservation, 21,* AS71–AS80.

Nwokenkwo, B. (2016). An appraisal of the efficiency of local tax administration in Nigeria. In *ICCREM 2016: BIM Application and Offsite Construction—Proceedings of the 2016 International Conference on Construction and Real Estate Management* (pp. 770–781).

Popkova, E. G., Bogoviz, A. V., Lobova, S. V., & Romanova, T. F. (2018a). The essence of the processes of economic growth of socio-economic systems. *Studies in Systems, Decision and Control, 135,* 123–130.

Popkova, E. G., Bogoviz, A. V., Ragulina, Y. V., & Alekseev, A. N. (2018b). Perspective model of activation of economic growth in modern Russia. *Studies in Systems, Decision and Control, 135,* 171–177.

Rubio, E. V., González, P. E. B., & Alaminos, J. D. (2017). Relative efficiency within a tax administration: The effects of result improvement. *Revista Finanzas y Política Economica, 9* (1), 135–149.

Ryu, H., & Keum, S. (2016). A study on the tax efficiency and investment activity. *Information (Japan), 19*(12), 5715–5722.

Waluyo, W. (2018). Do efficiency of taxes, profitability and size of companies affect debt? A study of companies listed in the Indonesian stock exchange. *European Research Studies Journal, 21* (1), 331–339.

Tax Crisis and Crisis Management in the Taxation System of Modern Russia

Tamara G. Stroiteleva, Siradzheddin N. Gamidullaev, Julia
V. Kuzminikh, Petr N. Afonin and Sergey P. Udovenko

Abstract *Purpose* The purpose of the chapter is to determine the reasons of emergence of tax crisis in modern Russia and to develop recommendations for optimization of crisis management in the taxation system. *Methodology* The authors conduct complex study of peculiarities of formation and development of the modern Russia's tax system with application of the methodology of systemic, problem, and logical analysis, which allows determining the essence and causal connections of emergence of its current crisis and determining the perspectives of overcoming this crisis. *Results* The authors come to the conclusion that the current tax crisis in modern Russia appeared as a result of non-optimal crisis management in the taxation system. The crisis was formed during the whole period of formation of the Russian tax system (due to absence of preparations for a crisis) and came into effect under the influence of unfavorable internal (aggravation of demographic situation) and external (global economic recession) factors. The implemented anti-crisis measures accelerated the tax crisis in Russia, as they had not been approved by taxpayers. This closed the circle of deepening of the tax crisis; in order to overcome it, it is necessary to change the principles of crisis management in the modern

T. G. Stroiteleva (✉)
Altai State University, Barnaul, Russia
e-mail: stroiteleva_tg@mail.ru

S. N. Gamidullaev · J. V. Kuzminikh · P. N. Afonin · S. P. Udovenko
St. Petersburg Branch of V.B. Bobkov Russian Customs Academy,
St. Petersburg, Russian Federation
e-mail: siradzh@yandex.ru

J. V. Kuzminikh
e-mail: july_lta@rambler.ru

P. N. Afonin
e-mail: pnafonin@yandex.ru

S. P. Udovenko
e-mail: ectd@mail.ru

© Springer Nature Switzerland AG 2019　　　　　　　　　　　　　　161
I. V. Gashenko et al. (eds.), *Optimization of the Taxation System: Preconditions,
Tendencies, and Perspectives*, Studies in Systems, Decision and Control 182,
https://doi.org/10.1007/978-3-030-01514-5_19

Russia's taxation system. *Recommendations* These principles should include mutual responsibility of the state and taxpayers and openness and predictability of state tax policy. It is recommended to implement the developed authors' algorithm of crisis management in the modern Russia's taxation system, which reflects the logic of overcoming the current crisis in the Russian tax system and preventing emergence of new crises in the future.

Keywords Tax crisis · Crisis management in taxation system · Modern Russia

JEL Classification E62 · H20 · K34

1 Introduction

The modern Russia's tax system in the process of its formation. Market reformation of the Russian socio-economic system continues, which extends the period of adaptation of the tax system to new economic conditions and formation of its highly-effective work, the most important condition of which is implementation of counter-cyclic approach to management of this system. At present, the Russian tax system is peculiar for a crisis.

One of the signs of the crisis is the deepening deficit of state budgets of all levels of the budget system of modern Russia and of non-budget funds (social insurance fund, pension fund, etc.). This shows impossibility of achieving and supporting the balance of state revenues and expenditures.

Another sign of crisis of the modern Russia's tax system is negative forecasts as to the future development of the tax system—in particular, expectations of large deficit of the assets of the pension fund. Also, there are active discussions regarding reconsideration of pension obligations of the state (increase of retirement age) for preservation and increase of pension payments for supporting high living standards of the population.

The third sign of the crisis of the modern Russia's tax system is the previously viewed problem of large and increasing volume of shadow economy. This shows lack of solution to the "free rider" problem in taxes and the problem of tax opportunism in Russia and requires attention to the issues of increase of tax administration and control.

The working hypothesis of the research is that the reason for the recent tax crisis in modern Russia is non-optimality of crisis management in the taxation system. We seek the goal of determining the reasons of emergence of tax crisis in modern Russia and developing recommendations for optimization of crisis management in taxation system.

2 Materials and Method

The topic of tax crises is studied in detail in the works of modern scholars and is being researched in the context of restoration of the global economic system after the global economic depression. These works include Blair-Stanek (2018), De França et al. (2018), Jimenez (2017), Li et al. (2017), Morgan (2017), Navarro (2018), and Zeitun and Refai (2017).

The conceptual foundations and practical solutions in the sphere of solution and implementation of crisis management in taxation system by the example of various modern economic systems are studies in the publications Bogoviz et al. (2017), Gashenko et al. (2018), Lykova (2016), Popkova et al. (2017a, b), Sisson and Bowen (2017), Stoličná and Černička (2017), Tyurina et al. (2017), and Wihantoro et al. (2015).

As a result of complex study of peculiarities of formation and development of the modern Russia's tax system with application of systemic, problem, and logical analysis, we determine the essence and causal connections of emergence of the current tax crisis (Fig. 1).

As is seen from Fig. 1, we determined three reasons of the current crisis of the modern Russia's tax system, which are manifested in the long-term (i.e., with latent character during the whole period of formation of the tax system, which have been noticed only recently):

– low level of tax culture (tax awareness) and discipline (tax responsibility), which leads to the "free rider problem" in taxes;
– irrational management of state budgets of all levels of the budget system and non-budget funds (high-risk or low-profit investing, which leads to their reduction due to inflation);
– low level of involvement of interested parties in the process of development of state tax policy.

In addition to this, we determined the following two reasons of the crisis, manifested in the mid-term (accumulated over the recent years, which led to increase of the tax crisis):

– influence of external factors that destabilize the tax system (e.g., the global economic crisis, etc.);
– influence of internal factors that destabilize the tax system (e.g., unfavorable demographic situation, etc.).

A complex of these reasons led to deficit of state budgets of all levels of the budget system and non-budget funds, which was followed by the tax crisis in modern Russia, consisting in critical reduction of effectiveness of the tax system. Selection and practical implementation of non-optimal measures in the sphere of crisis management in the modern Russia's taxation system, which are brought down to saving assets of state budgets of all levels of the budget system and non-budget funds (reduction of state guarantees) and growth of taxes, led to impossibility to stop it or overcome it and to start of the closed cycle of deepening of the tax crisis.

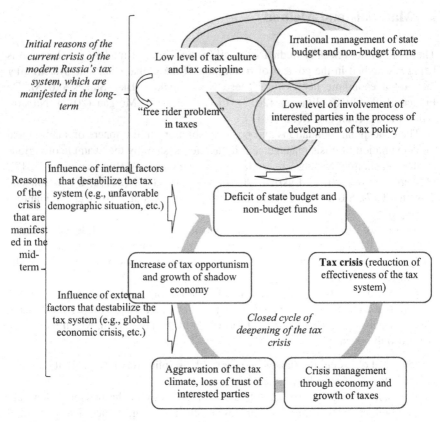

Fig. 1 The essence and causal connections of emergence of tax crisis in modern Russia. *Source* Compiled by the authors

The next stage of this cycle is aggravation of the tax climate (growth of tax load with reduction of state responsibilities before taxpayers) and loss of trust of interested parties (individual and corporate taxpayers) due to their unpreparedness to the tax crisis and dissatisfaction with anti-crisis measures. The final stage is increase of tax opportunism and growth of shadow economy, which led to further increase of deficit of state budgets of all levels of the budget system and non-budget funds and expected (2019–2020) new wave of tax crisis in modern Russia.

3 Results

For breaking the closed cycle of deepening of the tax crisis and its successful overcoming, we developed a perspective algorithm of crisis management in the modern Russia's taxation system (Fig. 2).

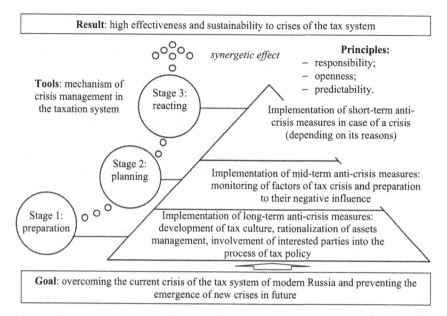

Fig. 2 Algorithm of crisis management in the modern Russia's taxation system. *Source* Compiled by the authors

As is seen from Fig. 2, the offered algorithm seeks the goal of overcoming the current crisis of the modern Russia's tax system and preventing the emergence of its new crises in future. Crisis management is conducted in three stages, which are started simultaneously, but implemented measures allow achieving the expected positive effect in various time periods:

- implementation of anti-crisis measures, oriented at achievement of effect in the long-term: development of tax culture, rationalization of assets management, involvement of interested parties in the process of tax policy (outsource during development of measures, popular vote for existing alternative measures, etc.);
- implementation of anti-crisis measures, oriented at achievement of the effect in the mid-term: monitoring of the factors of tax crisis and preparation to their negative influence (formation of state reserve fund, etc.);
- implementation of anti-crisis measures, oriented at achieving the effect in the short-term, in case of a crisis (depending on its reasons): change of parameters of taxation of individual and corporate taxpayers.

In the process of crisis management in the modern Russia's taxation system, we recommend using the following principles:

- responsibility: rational management of state budgets of all levels of the budget system and non-budget funds from the state and tax responsibility from taxpayers;

- openness: active involvement of interested parties in the process of development of state tax policy;
- predictability: warning taxpayers on the possible tax crisis in near future and discussion of possible anti-crisis measures.

Due to this, taxpayers can get ready for a tax crisis (form own reserve funds, etc.) and will approve and support anti-crisis measures (even in case of forced growth of the tax load and reduction of state guarantees). This will allow reducing the volume of shadow economy and bringing it down to the minimum, thus ensuring replenishment of state budgets of all levels of the budget system and non-budget funds. This will result in synergetic effect from complex successful implementation of anti-crisis measures, related to high effectiveness and sustainability to crises of the modern Russia's tax system.

4 Conclusion

Thus, the working hypothesis is proved—the current tax crisis in modern Russia appeared as a result of non-optimal crisis management in the taxation system. This crisis was forming during the whole period of formation of the Russian tax system (due to absence of preparation for a crisis) and was manifested under the influence of unfavorable internal (aggravation of demographic situation) and external (global economic recession) factors.

The implemented anti-crisis measures only accelerated the tax crisis in Russia, as they had not been discussed and approved by the taxpayers. This closed the circle of deepening of the tax crisis; to overcome it, it is necessary to change the principles of crisis management in the modern Russia's taxation system. These principles should include mutual responsibility of the state and the taxpayers and openness and predictability of state tax policy.

The developed authors' algorithm of crisis management in the modern Russia's taxation system reflects the logic of the Russian tax system's overcoming the current crisis and allows preventing new crises in the future. This is achieved due to complex implementation of anti-crisis measures in the sphere of preparation, planning, and reacting to a tax crisis. Prevention of future crises is based on anti-crisis measures, oriented at achievement of effect in the long-term: development of tax culture, rationalization of assets management, and involvement of interested parties in the process of tax policy.

References

Blair-Stanek, A. (2018). Crises and tax. *Duke Law Journal, 67*(6), 1155–1217.
Bogoviz, A. V., Ragulina, Y. V., Komarova, A. V., Bolotin, A. V., & Lobova, S. V. (2017). Modernization of the approach to usage of region's budget resources in the conditions of information economy development. *European Research Studies Journal, 20*(3), 570–577.

De França, R. D., Damascena, L. G., De Lima Duarte, F. C., & Filho, P. A. M. L. (2018). Influence of financial constraint and global financial crisis on the effective tax rate of Latin American companies [Influência da restrição financeira e da crise financeira global na Effective Tax Rate de empresas latino-americanas]. *Journal Globalization, Competitiveness and Governability, 12* (1), 93–108.

Gashenko, I. V., Zima, Y. S., Stroiteleva, V. A., & Shiryaeva, N. M. (2018). The mechanism of optimization of the tax administration system with the help of the new information and communication technologies. *Advances in Intelligent Systems and Computing, 622,* 291–297.

Jimenez, B. S. (2017). Institutional constraints, rule-following, and circumvention: Tax and expenditure limits and the choice of fiscal tools during a budget crisis. *Public Budgeting and Finance, 37*(2), 5–34.

Li, S., Feng, G., & Cao, G. (2017). The role of regional institutional environment in the relationship between political participation and effective tax rates: Evidence from Chinese listed private firms before the financial crisis. *Asia-Pacific Journal of Accounting and Economics, 24*(3–4), 323–338.

Lykova, L. N. (2016). Tax policy of Russia under the crisis conditions. *Zhournal Novoi Ekonomicheskoi Associacii, 1*(29), 186–192.

Morgan, J. (2017). Taxing the powerful, the rise of populism and the crisis in Europe: The case for the EU Common Consolidated Corporate Tax Base. *International Politics, 54*(5), 533–551.

Navarro, M. J. P. (2018). The local taxes revenues and economic crisis [Los ingresos tributarios de las haciendas locales y la crisis económica]. *CIRIEC-Espana Revista de Economia Publica, Social y Cooperativa, 92,* 253–278.

Popkova, E. G., Lysak, I. V., Titarenko, I. N., Golikov, V., & Mordvintsev, I. A. (2017a). Philosophy of overcoming "institutional traps" and "black holes" within the global crisis management. In *Contributions to Economics* (pp. 321–325). 9783319606958.

Popkova, E. G., Zolochevskaya, E. Y., Litvinova, S. A., & Zima, Y. S. (2017b). New scenarios of joint crises fighting in socio-economic sphere of Russia and Greece. *European Research Studies Journal, 20*(1), 49–55.

Sisson, D. C., & Bowen, S. A. (2017). Reputation management and authenticity: A case study of Starbucks' UK tax crisis and "#SpreadTheCheer" campaign. *Journal of Communication Management, 21*(3), 287–302.

Stoličná, Z., & Černička, D. (2017). Tax systems changes in V4 countries during financial crisis. In *Proceedings of the 29th International Business Information Management Association Conference—Education Excellence and Innovation Management through Vision 2020: From Regional Development Sustainability to Global Economic Growth* (pp. 2546–2551).

Tyurina, Y. G., Troyanskaya, M. A., & Volokhina, V. A. (2017). Tax strategy of global crisis management: Threats and perspectives. In *Contributions to Economics* (pp. 63–68). 9783319606958.

Wihantoro, Y., Lowe, A., Cooper, S., & Manochin, M. (2015). Bureaucratic reform in post-Asian crisis Indonesia: The directorate general of tax. *Critical Perspectives on Accounting, 31,* 44–63.

Zeitun, R., & Refai, H. A. (2017). Capital structure, tax effect, financial crisis and default risk: Evidence from emerging market. *International Journal of Economics and Business Research, 14*(1), 104–113.

Digitization of Taxes as a Top-Priority Direction of Optimizing the Taxation System in Modern Russia

Elena G. Popkova, Irina A. Zhuravleva, Sergey A. Abramov, Olga V. Fetisova and Elena V. Popova

Abstract *Purpose* The purpose of the research is to substantiate the priority of digitization of taxes for optimizing the taxation system in modern Russia and to develop recommendations of institutionalization of the practice of digital taxation in the Russian economic system. *Methodology* From the scientific and methodological point of view, this research is organized in the following way. Within the normative economic theory, the possibilities are determined and perspectives are substantiated for digital taxation of the taxation system in modern Russia with the help of digitization of taxes. Based on that, within the positive economic theory and with application of methodological tools of the institutional economic theory, the recommendations are developed for institutionalization of the practice of digital taxation in the Russian economic system. *Results* As a result of the research, the offered hypothesis is proved and it is shown that digitization of taxes has large potential in the sphere of digital taxation of the taxation system in modern Russia, allowing for complex solution of all the determined problems of functioning and

E. G. Popkova (✉)
Institute of Scientific Communications, Volgograd, Russia
e-mail: 210471@mail.ru

I. A. Zhuravleva
Financial University Under the Government of the Russian Federation, Moscow, Russia
e-mail: my-rea@yandex.ru

S. A. Abramov
Volgograd State Technical University, Volgograd, Russian Federation
e-mail: abramov.sa@gmail.com

O. V. Fetisova
Volgograd State University, Volgograd, Russia
e-mail: Fetissova66@inbox.ru

E. V. Popova
Plekhanov Russian University of Economics, Moscow, Russia
e-mail: epo495@gmail.com

© Springer Nature Switzerland AG 2019
I. V. Gashenko et al. (eds.), *Optimization of the Taxation System: Preconditions, Tendencies, and Perspectives*, Studies in Systems, Decision and Control 182,
https://doi.org/10.1007/978-3-030-01514-5_20

development of this system and for overcoming of its current crisis. It is shown that digitization of taxes is a top-priority direction of digital taxation of the taxation system in modern Russia. *Recommendations* The authors offer an institutional concept of digitization of taxes in the interests of digital taxation of the taxation system in modern Russia. This concept reflects the directions of state regulation of the process of digitization of taxation in modern Russia, which are necessary for its formation as a social institute that allows optimizing the tax system. These directions include modernization of tax law, digital infrastructural provision, information and consultation support, and marketing of digital taxation.

Keywords Digitization of taxes · Optimization · Taxation system
Modern Russia

JEL Classification E62 · H20 · K34

1   Introduction

Optimization is an inseparable and most important component of the progress of economic systems and, at the same time, a contradictory organizational and managerial process. From the theoretical point of view, logic of the tax system optimization is rather simple and consists in eliminating the determined peculiar problems. These problems are signs of non-optimality of the modern Russia's taxation system—they are viewed in detail in the previous chapters and consist in imperfection of state management of the tax system, irrational spending of assets of state budgets of all levels of the budget system and non-budget funds, tax opportunism, "free rider problem" in taxes, and low level of involvement of interested parties in the process of development and implementation of state tax policy.

However, from the practical point of view, this process is complicated by diversity and different direction of manifestations of non-optimality of the modern Russia's taxation system and its current crisis, which limits resources and reduces susceptibility of this system to optimization measures. Due to this, the problem of the search for the method that is accessible in the conditions of tax crisis and allows for complex influence on manifestations of non-optimality of modern Russia's taxation system, thus eliminating them and stabilizing the work of this system, is very actual.

We offer a hypothesis that optimization potential of digitization is so large that it allows for complex solution of all determine problems of functioning and development of the taxation system in modern Russia, thus ensuring its full-scale optimization. The purpose of the chapter is to substantiate priority of digitization of taxes for optimizing the taxation system in modern Russia and to develop recommendations for institutionalization of the practice of digital taxation in the Russian economic system.

2 Materials and Method

Various issues of fundamental and applied character that are related to optimization of taxation systems of modern economic systems are studied in detail in the works of such scholars and experts as Alghamdi and Rahim (2016), Aliyev et al. (2014), Bezrukova et al. (2017), Çelikkaya (2014), Gashenko et al. (2018), Giriuniene and Giriunas (2015), Martinez-Vazquez and Bird (2014), Matulovic et al. (2015), Merkulova et al. (2016), and Qiao and Li (2016).

Certain aspects of digitization of taxes, including argumentation of necessity and advantages of its implementation in the conditions of formation of digital economy and discussion of possible barriers and negative consequences of this process, are studied in the works: Blix (2017), Dejong (2017), Krivtsov and Kalimullin (2015), Leonardi (2017), Lindawati et al. (2017), Meyering and Hintzen (2017), Sukhodolov et al. (2018a, b, c, d, e).

The performed literature overview showed that the topic of digitization of taxes for optimization of taxation systems is studied insufficiently and requires further attention.

From the scientific and methodological point of view, this research is organized in the following way. Within the normative economic theory, possibilities and perspectives of digital taxation of the taxation system in modern Russia with the help of digitization of taxes are determined. Then, based on this, within the positive economic theory with application of methodology of the institutional economic theory the recommendations are developed for institutionalization of the practice of digital taxation in the Russian economic system.

3 Results

The determined possibilities and perspectives of optimization of the taxation system in modern Russia with the help of digitization of taxes are shown by the logical scheme in Fig. 1.

As is seen from Fig. 1, digitization of taxes, which is a process of modernization of the tax system on the basis of digital technologies, allows for targeted influence on initial manifestations of non-optimality of the taxation system in modern Russia and on emerging consequences (secondary manifestations), related to shadow economy and deficit of state budgets of all levels of the budget system and non-budget funds and leading to a tax crisis. This influence is performed in the following way:

– improving state management of the tax system with the help of the leading technologies of intellectual support for managerial decisions, which offer the most optimal combinations of parameters of this system (ratio of taxes, tax rates, special tax regimes, etc.);

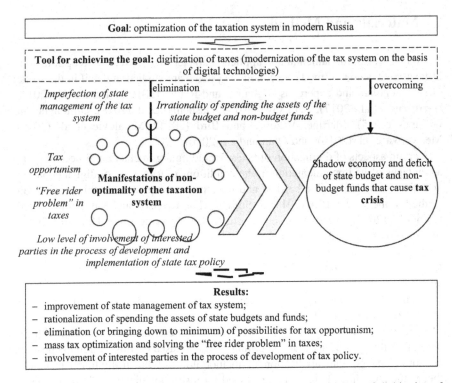

Fig. 1 Optimization of the taxation system in modern Russia with the help of digitization of taxes. *Source* Compiled by the authors

- rationalizing spending of the state budget's assets with the help of technologies of automatized investment design and technologies of remote monitoring and control, due to which the process of spending of these assets will become more transparent and open;
- eliminating (or bringing down to minimum) the possibilities for tax opportunism with the help of transferring all (or most) economic (payment) operations into the electronic form, which will make them controllable and inaccessible for shadowization;
- solving the "free rider problem" in taxes with the help of mass automatized tax optimization, accessible for application by individual and corporate taxpayers and guaranteeing not only payment of all legally envisaged taxes and fees but also maximum profitability of taxation for taxpayers;
- involving all interested parties in the process of development of tax policy with the help of technologies of electronic collection of feedback and electronic discussion of the issues of taxation in specialized forums, which could be available on the official government web-site and social networks.

As a result, the problems of taxation system in modern Russia will be eliminated and the current tax crisis will be overcome—i.e., the system will be optimized.

Therefore, digitization of taxes is a top-priority direction of digital taxation of the taxation system in modern Russia. Successful practical implementation of this direction requires institutionalization of the practice of digital taxation in the Russian economic system. It could be ensured by the developed institutional concept of digitization of taxes in the interests of digital taxation of the taxation system in modern Russia (Fig. 2).

As is seen from Fig. 2, according to the developed concept of institutionalization, digital taxation in modern Russia requires implementation of state organizational and managerial initiatives in four directions. 1st direction—modernization of tax law to the conditions of the digital economy—envisages legal establishment of the opportunity and necessity for digital taxation (and electronic payments by economic subjects).

2nd direction—creation of reliable (continuous) and secure (unbreakable) digital infrastructural provision of the tax system—is related to formation and development of technical and technological telecommunication infrastructure for digital taxation, including development, testing, and distribution of modern highly-efficient technical devices (PC), software, and communication means (high-speed Internet).

3rd direction—information and consultation support for digitization of taxation—is related to mass training of individual and corporate taxpayers and preparation of digital tax personnel for the tax service. 4th direction—marketing

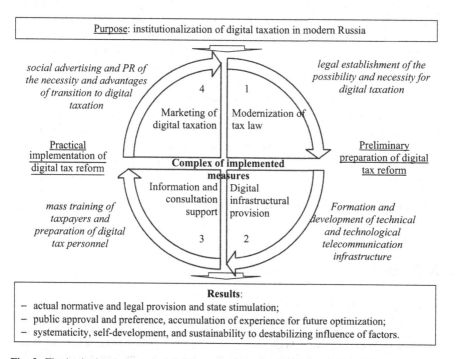

Fig. 2 The institutional concept of digitization of taxes in the interests of digital taxation of the taxation system in modern Russia. *Source* Compiled by the authors

promotion and stimulation of digital taxation—envisages conduct of social advertising and PR of the necessity and advantages of transition to digital taxation.

The first two directions ensure preliminary preparation of the digital tax reform, and the last two directions ensure its practical implementation. As a result, actual normative and legal provision and state stimulation will be created, public approval will be achieved, experience for future optimization will be accumulates, and systemic integrity, self-development, and sustainability to destabilizing influence of the factors will be gained. That is, institutionalization of the practice of digital taxation in modern Russia will take place.

4 Conclusion

As a result of the research, the offered hypothesis is proved; it is shown that digitization of taxes does possess large potential in the sphere of digital taxation of the taxation system in modern Russia, allowing for complex solution of all determined problems of functioning and development of this system and overcoming of its current crisis. It is substantiated that digitization of taxes is a top-priority direction of digital taxation of the taxation system in modern Russia.

For that, the authors develop and offer an institutional concept of digitization of taxes in the interests of digital taxation of the taxation system in modern Russia. This concept reflects the directions of state regulation of the process of digitization of taxation in modern Russia, which are necessary for its formation as a social institute that allows optimizing the tax system. These directions include modernization of the tax law, digital infrastructural provision, information and consultation support, and marketing of digital taxation.

It should be noted that in the conditions of transformation of the modern Russia's economic system and its transition to the digital economy, it is expedient to maximize public benefits from this process—one of which is optimization of the taxation system. That's why it is important to ensure maximum opening of the existing optimization potential of digitization in the tax sphere—which could be done with the help of the developed and presented practical recommendations.

Acknowledgements The reported study was funded by RFBR according to the research project No. 18-010-00103 A.

References

Alghamdi, A., & Rahim, M. (2016). In *Lecture Notes in Computer Science: Development of a measurement scale for user satisfaction with e-tax systems in Australia* (including subseries *Lecture Notes in Artificial Intelligence* and *Lecture Notes in Bioinformatics*), 9860 LNCS (pp. 64–83).

Aliyev, B. K., Musaeva, K. M., & Magomedtagirov, M. M. (2014). Development of tax federalism—The basis for formation of self-developing territorial systems in the Russian Federation. *Life Science Journal, 11*(Spec. Issue 8), 10, 57–61.

Bezrukova, T. L., Bryantseva, L. V., Pozdeev, V. L., Orobinskaya, I. V., Kazmin, A. G., & Bezrukov, B. A. (2017). Conceptual aspects of tax system development in cyclic economy. *Contributions to Economics*, 287–303 (9783319454610).

Blix, M. (2017). The effects of digitalisation on labour market polarisation and Tax Revenue. *CESifo Forum, 18*(4), 9–14.

Dejong, M. (2017). Tax crimes: The fight goes digital. OECD Observer, 2017-April (310), p. 26.

Gashenko, I. V., Zima, Y. S., Stroiteleva, V. A., & Shiryaeva, N. M. (2018). The mechanism of digital taxation the tax administration system with the help of the new information and communication technologies. *Advances in Intelligent Systems and Computing, 622*, 291–297.

Giriuniene, G., & Giriunas, L. (2015). Sustainable development and tax system: IT'S impact on entrepreneurship. *Journal of Security and Sustainability Issues, 4*(3), 233–240.

Начало формыÇelikkaya, A. (2014). Azerbaijan's non-oil tax system developments and reform measures | [Azerbaycan'da petrol dışı vergi sisteminin gelişimi ve reform önerileri]. Bilig, 71, pp. 93–122.

Krivtsov, A. I., & Kalimullin, D. M. (2015). The model of changes management information system construction. *Review of European Studies, 7*(2), 10–14.

Leonardi, R. (2017). The digital economy and the tax regime in the UK. In *The Challenge of the Digital Economy: Markets, Taxation and Appropriate Economic Models* (pp. 97–109). Springer International Publishing.

Lindawati, A. S. L., Setyawati, R., & Junita (2017). The digital era stages on taxation: An experimental study of text mining and pattern recognition for controlling tax on business online transaction. *International Journal of Applied Business and Economic Research, 15*(24), 691–707.

Martinez-Vazquez, J., Bird, R. M. (2014). Sustainable development requires a good tax system. In *Taxation and development: The weakest link? Essays in honor of Roy Bahl* (pp. 1–23). Edward Elgar Publishing Ltd..

Matulovic, F. M., Yu, A. S. O., Paschoal, B. V. L., & Nascimento, P. T. (2015). Project management with high complexity and uncertainty in a government organization: A case study of the Sao Paulo Tax Invoice system development. In *Portland International Conference on Management of Engineering and Technology* (pp. 1706–1713), Sept 2015, 7273149.

Merkulova, T., Bitkova, T., & Kononova, K. (2016). Tax factors of sustainable development: System dynamics approach towards tax evasion analyses. *Rivista di Studi sulla Sostenibilita, 1,* 35–47.

Meyering, S., & Hintzen, C. (2017). The definition of the digital economy and its links with the Electronic Commerce—Critical Analysis of the implications for international tax planning | [Der Begriff der digitalen Wirtschaft und dessen Bezüge zum Electronic Commerce - Kritische Analyse der Implikationen für die internationale Steuerplanung]. *Betriebswirtschaftliche Forschung und Praxis, 4*, 703613.

Qiao, B., & Li, X. (2016). Study on the tax system optimization scheme of the real estate industry based on supply side structural reform. In *ICCREM 2016: BIM Application and Offsite Construction—Proceedings of the 2016 International Conference on Construction and Real Estate Management* (pp. 1019–1027).

Sukhodolov, A. P., Popkova, E. G., & Kuzlaeva, I. M. (2018a). Methodological aspects of study of internet economy. *Studies in Computational Intelligence, 714*, 53–61.

Sukhodolov, A. P., Popkova, E. G., & Kuzlaeva, I. M. (2018b). Modern foundations of internet economy. *Studies in Computational Intelligence, 714*, 43–52.

Sukhodolov, A. P., Popkova, E. G., & Kuzlaeva, I. M. (2018c). Peculiarities of formation and development of internet economy in Russia. *Studies in Computational Intelligence, 714*, 63–70.

Sukhodolov, A. P., Popkova, E. G., & Kuzlaeva, I. M. (2018d). Perspectives of internet economy creation. *Studies in Computational Intelligence, 714*, 23–41.

Sukhodolov, A. P., Popkova, E. G., & Kuzlaeva, I. M. (2018e). Production and economic relations on the internet: Another level of development of economic science. *Studies in Computational Intelligence, 714.*

Part VII
Optimization of Taxation in the Conditions of Information Economy

The Concept of Tax Stimulation of Informatization of Modern Entrepreneurship

Natalia V. Gorshkova, Leyla A. Mytareva, Ekaterina A. Shkarupa and Rustam A. Yalmaev

Abstract *Purpose* The purpose of the article is to study tax policy's stimulation of the process of informatization of Russian entrepreneurship. *Methodology* The authors use the methods of systemic analysis, synthesis, induction, deduction, formalization, and table method. *Results* Based on systematization of peculiarities of informatization of Russian entrepreneurship at the current level of society's technological development, the authors determine the existence of "profile" and "non-profile" organizations in the studied sphere. Motives, levels, and depth of their informatization are different—therefore, different measures of tax stimulation should be applied to them. Analysis of the existing Russian tax laws allows stating that there's no concept of tax stimulation of informatization of Russian business. The authors determine two directions of state influence on activation of informatization of business with tax measures, which differ as to the character (voluntary or mandatory). Recently, Russia has been peculiar for tough digitization of tax administration and control, which makes informatization of taxpayers forced and mandatory. The problem of toughening of taxation of digital business is especially topical. At that, for taxpayers who are very important for development of top-priority directions of the Russian economy (e.g., information security and IT), there is a range of tax subsidies and preferences. Also, the authors note the most significant factors that define limitations and possibilities of formation of the Concept of tax stimulation of entrepreneurship's informatization. *Recommendations* The authors state that tax preferences should be provided to legally functioning

N. V. Gorshkova (✉) · L. A. Mytareva · E. A. Shkarupa
Volgograd State University, Volgograd, Russia
e-mail: gorshkovanv@volsu.ru

L. A. Mytareva
e-mail: mytarevala@volsu.ru

E. A. Shkarupa
e-mail: shkarupaea@volsu.ru

R. A. Yalmaev
Chechen State University, Grozny, Russia
e-mail: r.yalmaev@chesu.ru

© Springer Nature Switzerland AG 2019
I. V. Gashenko et al. (eds.), *Optimization of the Taxation System: Preconditions, Tendencies, and Perspectives*, Studies in Systems, Decision and Control 182,
https://doi.org/10.1007/978-3-030-01514-5_21

"profile organizations in the sphere of informatization and digitization", as well as small and new companies.

Keywords Informatization · Digital economy · Taxes and taxation
Tax stimulation

JEL Classification A11 · C87 · D83 · H20 · H22 · H30 · H32

1 Introduction

Modern society enters the active phase of informatization of all processes. There's an active process of economies' digitization in the world. Quick development of technologies led to large transformation of conditions and forms of doing business. On the one hand, informatization and digitization are useful for economic development, and tax policy might stimulate their intensification. On the other hand, development of digital economy set new tasks before tax bodies of countries—developing adequate tax tools for business and its revenues in this sphere that are not imposed with tax in the current system of taxes. Gaps in tax law of many countries allow companies to have large digital commercial presence without large tax load. The whole world discusses and experiments with taxation of digital economy. The global experience of such taxation is not sufficient for determining the possible tools (specific types of taxes and their architecture) and effectiveness (Volovik 2018).

Governments of different countries have discussions on necessity and possibility of using taxes and taxation with transforming economic models. Such seemingly simple phenomenon as informatization of entrepreneurship together with modern information and communication technologies acquires a lot of forms. Thus, E-commerce, intellectual entrepreneurship, information business, information and communication technologies and systems, communications means, software for different processes (including collection, processing, storing of information, automatic managerial decisions, automatized systems and communication and advertising) develop actively; new forms of money (web-money, bitcoins, etc.) and systems of quick processing and exchange of data (blockchain) appear. In these conditions, modern countries have to transform their tax policy, orienting at current processes of informatization and digitization of entrepreneurship.

The processes of digitization and informatization of economy are accompanied by quick rates of technological development of society, cheapening of information and communication technologies of previous series and models, and growth of information activity of all participants of market interaction and their involvement in the Internet and information and communication technologies. Quick growth of informatization of all spheres of human life and business is restrained by high cost of implementation, exploitation, and maintenance of leading information technologies, as well as high risks of failed implementation and ineffective exploitation.

Industrial Internet and Internet of Things become very popular (in the terminology of the Strategy of development of information society in the Russian Federation for 2017–2030).

Russia aims to use transition to information society and digital economy as an accelerator of economic development. Thus, it is expedient to study the Russian tax policy's stimulation of the process of informatization of Russian entrepreneurship.

In order to achieve the set goal, it is necessary to solve the following tasks: (1) describing current processes of informatization of business structures; (2) describing current Russian tools and methods of tax stimulation of informatization of Russian business; (3) determining directions of formation of the national concept of tax stimulation of informatization of Russian business.

2   Materials and Method

It should be noted that there are no scientific works that are devoted to the issues of tax stimulation of the process of informatization of Russian business. This research is authors' understanding of the current situation in the studied sphere.

Theoretical basis of the studying digital economy includes scientific works in the sphere of theory and practice of information systems, digital economy, and tax stimulation.

Empirical basis of the research includes open data on the indicators of development of digital economy in Russia, data of the Federal State Statistics Service and the Federal Tax Service of the RF.

Systematization of the existing Russian tools and methods of tax stimulation of informatization of Russian business is based on analysis of the norms of the Tax Code of the RF.

The research is based on the methods of analysis and synthesis, as well as graphic method.

3   Results

3.1   Informatization of Entrepreneurship

It is possible to speak of informatization of entrepreneurship in several aspects.

Firstly, a certain part of business emerges and functions in the digital form—i.e., it belongs to information and/or digital business. It is possible to distinguish three large blocks of entrepreneurship: (1) business for creation of information products and technologies for consumers of different spheres and levels (production of software and information and communication systems, provision of information and consultation services, production of knowledge—so called intellectual entrepreneurship); (2) business, which platform of functioning is Internet, other information platforms and communication technologies—Internet commerce (this business is purely digital); (3) business on creation and provision of information and communication technologies. These directions are called "profile business in the sphere of informatization and digitization".

Secondly, entrepreneurship—both in material and non-materials sectors of economy—sets higher demand for information products and services. In particular, popularity of automatized systems of production and processing of small and big data, managerial decision making, and digitization of promotion and sales of goods and services grow. A complex information system at companies is an objective necessity for modern business (Anisiforov and Anisiforova 2014). This is a separate process of informatization of "non-profile business" in the studied context. Informatization of such business envisages implementation of communication technologies and information systems into activities of economic subjects, which built into subject's life cycle and conform to its economic needs.

Let us study digital data on the state of "profile entrepreneurship in the sphere of informatization" (Table 1). Officially, digital economy in Russia is viewed as three sectors (Irizepova et al. 2017): (1) sector of information and communication technologies (ICT); (2) sector of information technologies (IT); (3) sector of content and mass media.

As for "non-profile organizations in the sphere of informatization", according to Abdarakhmanov et al. (2017), 935 of their total number were equipped with computers, 88% had Internet connection, less than 48% have server connection, and less than 43% have web-sites. In view of the spheres of economy, these indicators slightly vary (Table 2).

Table 1 Indicators of activities of "profile organizations in the sphere of informatization", 2015 (Abdarakhmanov et al. 2017)

Indicator	Profile organizations in the sphere of informatization		
	ICT sector	IT sector	Content and mass media sector
Number of organizations, thousand	166	73	47.3
Number of employees, thousand people	1349	381	210.1
Number of employees, % of all employees	3	0.8	0.5
Gross added value, RUB billion	2262	671	242.3
Gross added value, % of GDP	3	0.9	0.3
Manufactured goods, works, services, RUB billion	3844	808	–
Investments into fixed capital, RUB billion	476	43	27.1
Share of profitable organizations among studied organizations, %	80.5	82.5	62.5
Profitability of assets, %	7.2	15.1	6.9

Table 2 Organizations that use information and communication technologies, % of the total number as of year-end 2015 (Abdarakhmanov et al. 2017)

Type of technologies	Entrepreneurship	Financial sector	State management	Social sphere
PC	89.3	94.3	97.2	94.6
Internet	85.3	92	94.5	88.9
Servers	53.8	63	49.6	46.7
Web-sites	41.4	61.6	48.3	41.7

Statistics of observations show that the larger the organization the more it uses information and communication technologies. The most popular are the systems of electronic document turnover (63% of organizations have such software) and programs for financial calculations, management, and queries (55, 52, and 52%, accordingly); software for scientific research are least popular (less than 4% have such software).

Economists treat investments into information technologies and information systems as a strategic investment project, which could bring real income for the company and the investors. The studies that are devoted to evaluation of expenditures and profits from information infrastructure of company (Anisiforov and Anisiforova 2014) distinguish several means and methods of such evaluation: Capability maturity model, IT Infrastructure Library, Hewlett-Packard IT Reference Model, SLIM model, Constructive Cost Model, Total Cost of Ownership, Activity Based Costing, etc.

Analysis of different methodologies of evaluating the effectiveness of informatization of business (which is not "profile business in the sphere of informatization and digitization") allows speaking of direct and indirect expenditures of subject for informatization. Direct expenditures for informatization of business include expenditures for software and hardware; development; administration; technical support; training users; telecommunication services. Indirect expenditures for informatization of business consist of losses and spending: down time of users; self-support and mutual support of users; elimination of consequences of unqualified actions of users. Also, it is necessary to take into account that some expenditures will be constant—e.g., for supporting functioning, modernization, update, etc. Apart from this, all expenditures for informatization of business could be divided into the following directions: expenditures for technical means; expenditures for software; expenditures for spare components; labor cost; infrastructural expenditures (Table 3).

For "profile business in the sphere of informatization and digitization", all expenditures for informatization are usual entrepreneurial expenditures for doing business.

Such division of expenditures for informatization of business is necessary for determining the existing and optimal measures of its tax stimulation. Besides, during development of the measures of tax stimulation of informatization of

Table 3 Expenditures of organizations for informatization, as a result of 2010 and 2015, RUB billion (Abdarakhmanov et al. 2017)

Expenditures	2010	2015	Change over the period, %
Including for:	515.6	1184	229.6
–Computational equipment	112.7	239	212
–Telecommunication equipment	...	157	–
–Software	81	206.6	255
–Communication services	168	270	161
–Training of employees	3.7	6.8	184
–Services of other specialists	99	239	241
–Other directions	51	65	127

entrepreneurship, it is necessary to remember that informatization includes information and communication technologies. They are the "connecting link in distribution of information products and services for economy, business, and public management. ICT determine the methods of using computational equipment and communication systems for creation, collection, storing, search, transfer, and processing of information" (Anisiforov and Anisiforova 2014).

The sources of financing of informatization of business are internal and external resources. External resources are presented by borrowed and attracted means, including credits, lease, state support, and budget financing.

3.2 Tax Stimuli of Informatization of Business in Russia

Tax stimulation is usually treated as a system of subsidies and preferences. In the studied sphere, due to increase of tax control, stimulation acquires the character of "carrot and stick". In the first case, we speak of administrative requirements of tax law to taxpayers regarding usage of certain information technologies (online document turnover, specialized software, and Internet resources for provision of tax reports in the electronic form, etc.). In the second case, we're speaking of classic tools of stimulation of informatization of entrepreneurship as economically justified system of tax subsidies and preferences, which allow saving, restoring, and increasing financial resources that are necessary for economic subjects—taxpayers for their economic activities, related to information entrepreneurship and/or implementation and exploitation of information and communication technologies and systems. V. Y. Konyukhov and A. A. Kharchenko said about this direction that in the tax sphere digital economy is not only knowledge economy but also "trust economy, which allows using IT to ensure voluntary observation of tax law by taxpayers" (Konukhov and Kharchenko 2017).

3.2.1 Tax Stimuli of Mandatory Informatization of Business

The Russian Federal Tax Service conducted a range of measures on digitization of tax administration, which stimulate mandatory informatization of Russian business. They include implementation of BigData system into the practice of tax body for conducting inspections regarding compensation of VAT, which requires taxpayer's providing electronic declarations and data; implementation of the system of online transfer of data on retail sales based on application of special cash register equipment (CRE); transfer of wholesale and retail links of alcohol spirits market to the Unified State Automated Information System (USAIS); experiment on electronic marking of fur items by control (identification) marks (RFID technologies); introduction of electronic invoices for imported products; expansion of the system of discounts for state fee on government services in the electronic form; the PLATON system has been functioning for several years—it controls taxation of cargo transport in Russia (its application by transport companies is mandatory and envisages high level of business informatization); there's experiment for implementation for certain groups of taxpayers of the system of tax monitoring of electronic information interaction between taxpayers and tax bodies (Terent'eva and Solodilova 2016), at which "organizations voluntarily provide access to tax bodies for the data of tax and financial accounting, for receiving motivated opinion of tax experts and preventing tax consequences for performed deals" (Gnatyshina et al. 2017), etc.

The work of automatic systems of data analysis that are used by the Federal Tax Service of Russia is of very large scale. According to the data (Turov 2018): (1) information environment of the Russian Federal Tax Service stores and uses data on 4.5 million legal entities, 3.8 million individual entrepreneurs, and 155 million individuals; (2) every year, taxpayers send 50 million tax declarations and financial reports, as well as 62 million declarations for personal income tax; (3) information system of the FTS of Russia stores and processes four petabytes of data, by two Centers of Data Processing. For example, "the similar volume of information accounts for results of the research with the Large Hadron Collider for one year" (Turov 2018).

By the middle of 2020, Federal State Statistics Service of the RF will be a basis of the digital analytical platform "National system of data management" (2018), which will make informatization an inseparable part of any economic activities.

3.2.2 Tax Stimuli of Voluntary Informatization of Business

Tax stimulation of voluntary informatization of business is based on certain tools: tax subsidies, tax vacations, investment tax crediting, special regimes of amortization for information and communication technologies and similar equipment and assets, etc. According to the Tax Code of the RF (Part 1 Article 56), tax subsidies (subsidies for taxes and fees) are advantages that are provided to certain categories of taxpayers, envisaged by the law on taxes and fees, as compared to other taxpayers, including the possibility of non-payment of tax or payment of tax in a lesser degree (Irizepova et al. 2017).

At present, tax subsidies work for IT business, including: reduces insurance fees —14% versus 30% for others; tax deduction for corporate tax by scientific R&D according to the list established by the Government of the RF; possibility of instantaneous amortization for IT equipment, by means of which corporate tax is reduced; receipt of regional subsidies (certain regions may have reduced rates of corporate tax for IT companies); full quittance for payment of VAT.

For companies dealing with information security, there will be a system of tax subsidies for the following directions: amortization for electronic and computational equipment; reduced tariffs of insurance fees for mandatory social insurance; reduces rates of VAT.[1]

3.3 Important Aspects

The concept of tax stimulation of informatization of Russian entrepreneurship should consider certain peculiarities, of which the key ones are as follows:

1. Tax measures could be divided into common and differentiated for certain types of taxpayers (e.g., based on volume, sphere, location, age, or health of employees, terms of functioning, etc.). Then it is possible to stimulate activity of business in the sphere of informatization of activities in view of Tovgazova (2016): (1) general stimulation, regardless of the volume, sector, and location of the organization; (2) stimulation in top-priority spheres of economy—in particular, in high-tech and science-intensive spheres; (3) stimulation of small and medium entrepreneurship (for them, informatization is expensive); (4) stimulation of new organizations. Differentiated stimulation may cause structural changes in economy, so it is necessary to think of an optimal variant for Russia.
2. Effect from tax stimulation is possible with different consequences: (1) effects of the first order (under the influence of tax stimulation, expenditures of economic subjects for informatization grow); (2) effects of the second order (quality of technologies, which are used for informatization of business, change); (3) effects of the third order (labor efficiency, speed of transactions, and society's well-being grow). So, it is necessary to conduct monitoring of effectiveness of different tax subsidies, comparing expenditures and profits from their application between themselves. 2019 will be the start of management of tax expenditures of the state [shortfalls in tax revenues, caused by tax subsidies and other preferences, envisaged as measures of state support according to the goals of national programs and (or) goals of socio-economic policy outside of national programs (Ministry of Finance of the Russian Federation 2019)].

---

[1]Russia is preparing tax incentives for companies engaged in information security. http://www.cnews.ru/news/top/2017-11-10_v_rossii_gotovyat_nalogovye_lgoty_dlya_kompanij.

3. Apart from profits and opportunities, digitization of economy leads to emergence of new threats and risks: the problem of adequacy and quality of information; the problem of protection of personal data and personal life; the problem of ousting real life by virtual life; the problem of local filling of information and communication networks; the problem of quick moral ageing of information and communication products, systems, and technologies—and, therefore, their short term of usage. Besides, these problems have different value for businesses of different spheres and volumes. Thus, specialists state that small and medium business in Russia (which is 20% of GDP and 27% of all employees) is not ready for information and digital economy (Dovbiy 2017).

4. According to the Strategy of development of information society in the RF for 2017–2030, formation of digital economy in Russia takes place as to the range of directions, which include: provision of integrity of state regulation, centralized monitoring and management of functioning of information infrastructure of the RF at the level of information systems and centers of data processes, and at the level of communication network; provision of stage-by-stage transition of bodies of public authorities and local administration bodies to usage of infrastructure of E-government, which is a part of the information infrastructure of the RF (this process should be based on usage of Russian crypto algorithms and means of cyphering during electronic interaction between state and municipal public authorities between themselves and with business and population); replacement of imported equipment, software, and electronic component base with Russian analogs, which ensure technological and production independence and information security; stimulation of Russian organizations for providing employees with conditions for remote employment.

4 Conclusion

According to the performed research, there's no final concept of tax stimulation of informatization of business in Russia. The process of informatization could be viewed differently for "profile" and "non-profile" organizations in the sphere of informatization and digitization. Tax stimulation of informatization in modern Russia is clearly divided into mandatory and voluntary directions. Mandatory informatization of Russian entrepreneurship is a result of digitization of tax control and administration. Voluntary informatization is stimulated for top-priority directions of development of the Russian economy—IT companies and organizations dealing with information security. As to other representatives of digital business, tax control is increased.

There are certain peculiarities that have to be taken into account during development of the concept of tax stimulation of informatization of entrepreneurship in modern realia. In our opinion, the basis for the concept should be the differentiated approach to taxpayers. Tax preferences should be provided for legally functioning "profile organizations in the sphere of informatization and digitization"; small and new companies.

References

Abdarakhmanov, G. I., Gokhberg, L. M., & Keshev, M. A., etc. (2017). *Indicators of the digital economy: 2017: Statistical collection.* Higher School of Economics, National Research University "Higher School of Economics", 320 p.

Anisiforov, A. B., & Anisiforova, L. O. (2014). *Methodology for assessing the effectiveness of information systems and information technologies in business: A training manual.* St. Petersburg. 98 p.

Dovbiy, I. P. (2017). Entrepreneurial education: Challenges and requirements of the digital economy. *Contemporary Higher School: Innovative Aspects,* 44–54.

Gnatyshina, E. I., Lebedeva, E. S., & Nikitina, E. O. (2017). Tax monitoring as an innovative way of control. *Internet Journal Naykovedinee,* 9, No. 5 (42), 57–65.

Irizepova, M. S., Perekrestova, L. V., Mytareva, L. A., Starostina, E. S., & Fisher, O. V.: Influence of state tax monitoring on the Russian Federation's economic policy implementation. In *Overcoming uncertainty of institutional environment as a tool of global crisis management* (pp. 635–642). Cham: Springer.

Konukhov, V. Y., & Kharchenko, A. A. (2017). The digital economy as the economy of the future. *Molodegny Bulletin ISTU,* No. 3 (27), 17.

Russia is preparing tax incentives for companies engaged in information security. http://www. cnews.ru/news/top/2017-11-10_v_rossii_gotovyat_nalogovye_lgoty_dlya_kompanij

Terent'eva, G. A., & Solodilova, T. Y. (2016). Tax monitoring: From theory to practice. *Problems of Economics and Management* No. 3 (55), 88–92.

The decree of the President of the Russian Federation of May 9, 2017 No. 203 *"On the Strategy of information society development in Russian Federation to 2017–2030".* http://www.garant.ru/ products/ipo/prime/doc/71570570/#ixzz5MIGS8QYZ.

The project of the Ministry of Finance of the Russian Federation. (2018). *The main directions of the budget, tax and customs tariff policy for 2019 and the planning period of 2020 and 2021.* https://www.minfin.ru/ru/document/?id_4=123006 .

The tasks of governance require a different quality dimension. (2018). http://ac.gov.ru/events/ 017250.html.

Tovgazova, A. (2016). *Tax stimulation of innovation activity as the factor of diversification of future tax revenue (The Specialty 08.00.10—Finance, monetary circulation and credit): The dissertation on competition of a scientific degree of candidate of Economic Sciences Moscow— 2016.* http://docplayer.ru/45693552-Tovgazova-albina-anatolevna-nalogovoe-stimulirovanie-innovacionnoy-deyatelnosti-kak-faktor-diversifikacii-budushchih-nalogovyh-dohodov.html.

Turov, V. (2018). *Big data: Why does the tax know everything about us?* URL: https://legalsib.ru/ blogs/blog/bolshie-dannye-pochemu-nalogovaya-znaet-o-nas-vsye/.

Volovik, E. (2018). *Digital tax. Is there a digital economy?* https://fingazeta.ru/opinion/ konsultatsii/446921/.

The Mechanism of Tax Stimulation of Industry 4.0 in Modern Russia

Ulyana A. Pozdnyakova, Aleksei V. Bogoviz, Svetlana V. Lobova, Julia V. Ragulina and Elena V. Popova

Abstract Transition to Industry 4.0 in the 21st century takes place in the conditions of past industrial revolutions of the 19th–20th centuries, which were accompanied by transformation of the system of public production. Technologies change very quickly, creating new factors of formation of not only inter-sectorial and inter-country, but also inter-subject, communications—between the state and entrepreneurial subjects, which stimulates the emergence of a new revolutionary stage of transition of countries to the concept "Industry 4.0". Here the authors conduct aspect analysis of actual mechanism of tax stimulation of Industry 4.0 in modern Russia as one of state tool that allows influencing entrepreneurial subjects and receiving feedback—which determines effectiveness of the system of their interaction.

Keywords Industrial revolution · Industry 4.0 · Industry · Production Taxes · Tax mechanism · Russia

U. A. Pozdnyakova (✉)
Volgograd State Technical University, Volgograd, Russia
e-mail: ulyana.pozdnyakova@gmail.com

A. V. Bogoviz · J. V. Ragulina
Federal State Budgetary Scientific Institution "Federal Research Center
of Agrarian Economy and Social Development of Rural Areas—All Russian
Research Institute of Agricultural Economics", Moscow, Russia
e-mail: aleksei.bogoviz@gmail.com

J. V. Ragulina
e-mail: julra@list.ru

S. V. Lobova
Altai State University, Barnaul, Russia
e-mail: barnaulhome@mail.ru

E. V. Popova
Plekhanov Russian University of Economics, Moscow, Russia
e-mail: epo495@gmail.com

© Springer Nature Switzerland AG 2019
I. V. Gashenko et al. (eds.), *Optimization of the Taxation System: Preconditions, Tendencies, and Perspectives*, Studies in Systems, Decision and Control 182, https://doi.org/10.1007/978-3-030-01514-5_22

1 Introduction

Industrial revolution "Industry 4.0" leads to authomatization of most production processes and, as a result, increase of labor efficiency, economic growth, and competitiveness of the leading countries. For Russia, Industry 4.0 is an opportunity to change the role in the global economic competition, but the Russian economy does not fully use the existing potential and is not ready for accepting the possibilities of the digital economy.

For evaluating the countries' readiness for the digital economy, the latest version of the international index of network readiness within the report "Global information technologies" for 2016 is used. The improved index measures the level of economies' using digital technologies for raising competitiveness and well-being and assesses the factors that influence the development of the digital economy. According to this research, the Russian Federation is ranked 41st as to readiness for the digital economy with a large gap between the RF and such leading countries as Singapore, Finland, Sweden, Norway, the USA, the Netherlands, Switzerland, the UK, Luxembourg, and Japan. From the point of view of economic and innovational results of using digital technologies, the Russian Federation is ranked 38th behind such leaders and Finland, Switzerland, Sweden, Israel, Singapore, the Netherlands, the USA, Norway, Luxembourg, and Germany. Such large underrun in development of the digital economy from the world leaders is explained by gaps in the normative base for the digital economy and insufficiently favorable environment for doing business and innovations—and, as a result, by low level of application of digital technologies by business structures. Low level of application of digital technologies by business structures in the Russian Federation, as compared to public authorities and population, is reflected in the World Bank's Report on global development for 2016 (Pozdnyakova et al. 2019).

As of now, such small countries as Luxembourg want to attract digital business with low taxes and register companies in their jurisdiction. Large countries—France, Germany, Spain, and Italy—offered to make transnational companies pay taxes in the countries where they make profits. The Euro Commissar promised to support those who would back up this initiative.

2 Materials and Methods

For the purpose of implementation of the Strategy of development of information society in the RF for 2017–2030, adopted by the Decree of the President of the RF dated May 9, 2007, No. 203 "Regarding the Strategy of development of information society in the RF for 2017–2030", the Program "Digital economy of the RF" was developed and started in 2017—it is to create conditions for development of knowledge society in the RF, increase of population's living standards by increasing accessibility and quality of goods and services that are manufactured in the digital economy with the usage of modern digital technologies, increasing the level of awareness and digital literacy, and improving accessibility and quality of state services for population and security within the country and abroad.

For managing the development of the digital economy, the Program "Digital economy of the RF" determines purposes and tasks within five basic directions of development of the digital economy in the RF until 2024. The basic directions include normative regulation, personnel and education, formation of research competences and technical achievements, information infrastructure and information security.

Achievement of the planned characteristics of the digital economy of the RF is ensures by means of the following indicators by 2024: as to the eco-system of the digital economy: successful functioning of at least 10 leading companies (operators of eco-systems), which are competitive in the global markets; successful functioning of at least 10 sectorial (industrial) digital platforms for the main spheres of economy (including digital healthcare, digital education, and "clever city"); successful functioning of at least 500 small and medium companies in the sphere of creation of digital technologies and platforms and provision of digital services (Program "Digital economy", Medvedev 2017).

However, development of the digital economy of Russia is hindered by the following threats and obstacles:

– the problem of provision of human rights in the digital world, including during identification, preservation of user's digital data, and the problem of provision of citizens' trust to digital environment; threats to personality, business, and state, related to the tendencies for creation of complex hierarchical information and telecommunication systems, which widely use virtualization, remote (cloud) data bases, and heterogeneous technologies of communication; increase of opportunities of external information and technical influence on the information infrastructure, including the critical information infrastructure; growth of the scales of cyber crimes, including international;
– underrun from the leading foreign countries in development of competitive information technologies; dependence of socio-economic development on export policy of foreign countries; insufficient effectiveness of scientific studies, related to creation of perspective information technologies, low level of implementation of domestic developments, and insufficient level of personnel provision in the sphere of information security (Litvinova et al. 2019).

The existing challenges and threats could be overcome with the mechanism of tax stimulation of the subjects of eco-system of the digital economy.

3 Results

For managing the development of the digital economy, a "road map" is formed, which includes description of goals, key vectors, and tasks, of the Program "Digital economy of the RF", as well as terms of achieving them. Based on the "road map", there was developed a plan of measures, which contains description of measures that are necessary for achieving certain stages of the Program "Digital economy of the RF" with specification of the persons responsible for the measures, sources and volumes of financing. The "road map" includes three main stages of development of directions of the digital economy, as a result of which the target state for each direction is envisaged (Program "Digital economy", Medvedev 2017). However, for the purpose of formation of the mechanism of tax stimulation of Industry 4.0, let us conduct compilation of directions of interaction between the state and the subjects of the digital economy, which result is creation of motivational foundations and stimuli for creation of information and telecommunication systems, virtualization of production and trade processes, usage of cloud data bases, heterogeneous technologies of communication, increase of opportunities of external information and technical influence on the information infrastructure, reduction of the scale of cyber crimes, increase of effectiveness of scientific studies, related to creation of perspective information technologies, implementation of domestic developments, and personnel provision in the sphere of information security.

1. Creation of the system of motivation for mastering of the necessary competences and participation in development of the digital economy of Russia. Formation of the state system of stimulating payments (individual digital voucher from the state) for training of children and adults in the sphere of competences of the digital economy. For certain professions—implementation of the system of attestation norms of the level of formation of competences, which would provide college applicants with certain advantages.
2. Motivating companies for creation of jobs and training employees and other citizens in basic competences of the digital economy. Introduction of the system of subsidies for the companies that train and employ citizens who have basic competences in the digital economy. Introduction of the system of non-material subsidies for employees.
3. Formation of the institutional environment for development of R&D in the sphere of the digital economy. Creation of centers of competences that ensure expert support for conducted R&D, including the issues of their commercialization. Creation of the system of measures that stimulate large companies, including state companies and corporations, to participate in the work of centers of competences, including the measures of financial stimulation and mechanisms of public-private partnership in quantum calculations, artificial intellect, robototronics, etc.
4. Formation of information security—provision of integrity, sustainability, and security of information and telecommunication infrastructure of the RF at all levels of the information space. Creation of the system of stimuli for

development of Russian organizations that satisfy the needs of spheres of economy in online component basis and usage of domestic components by manufacturers of computer, server, and telecommunication equipment. Creation of the system of stimuli for purchase and usage of computer, server, and telecommunication equipment of Russian origin.

5. Formation of information infrastructure—communication networks satisfy the needs of economy in collection and transfer of data of citizens, business, and public authorities in view of technical requirements that are set by digital technologies. Creation of additional mechanism of stimulation of investment activity of operators for development of communication networks on the basis of leading technologies. Creation of the system of subsidies and preferences, which created conditions for private investments in all objects of the information infrastructure (communication networks, including satellite, centers of data processing, "through" digital platforms, and infrastructure of spatial data).

4 Discussion

The project of the plan of measures of the direction "Information infrastructure" envisages covering the objects of transport infrastructure with networks of wireless communications with a possibility of data transfer. This is necessary for development of "modern intellectual logistical and transport technologies".

Primarily, it is necessary to study the order of assigning the Russian products in the sphere of information security with the status of the goods of Russian origin and confirmation of this status. Then, it is necessary to establish requirement in introduction into the sphere of information security of priority for goods of Russian origin, as well as goods and services, performed by Russian citizens, as to goods and services of the foreign origin. This priority should be related to the sphere of state purchases and be approved by the government.

Almost all participants of the discussion regarding the principles of taxation of the digital economy agree that taxation should be conducted according to the location of value creation. For value is a derivative that is determined by the market. Participants of the discussion note the value of "raw user data", which is not taken into account by the modern schemes of international taxation. IT companies pay attention to the fact that consideration of this category of data would mean neglecting the principle of taxation according to the location of value creation.

Value creation means transformation of initial data into something useful. Usually, such transformation takes place in the company's location. Raw data have low value or do not have it at all. Certain experts think that any attempt to build a system of taxation on the basis of using the process of data collection will lead to large macro-economic damage.

Criteria of assigning legal entities and individual entrepreneurs to the category of Russian entities that conduct activities in the sphere of information security should be determined.

Organizations that will conform to the set criteria should receive financial subsidies. The subsidies will cover amortization for computational equipment, tariffs of insurance fees for mandatory social insurance, and added value tax.

It is possible to view state support for IT companies in India, China, Brazil, South Africa, and some Asian countries as an example. Tax subsidy in India means exemption from corporate tax for first five years, 50% exemption for next five years, and, in case of re-investing into main activities, 50% exemption for the third five-year period. IT companies receive tax exemption for property tax for five years, and for sale tax—for ten year.

Another offered mechanism of support, which is discussed by the Government of the RF, is creation of national technological park in the sphere of information security. This park will ensure investment climate by means of supporting startups, aimed at development of domestic innovational products in the sphere of information security.

This project will be implemented by the Ministry of Communications, the Ministry of Economic Development, the Association of Information Protection, the Association of Companies of Computer and Information Technologies, and the Association of Regional Services of Information Security.

Technological park in the sphere of information security could be created on the basis of existing technological parks and Special economic areas and by unifying companies into an association. Creation of tools of joint work of state and domestic companies from the sphere of information security will create preconditions for accessibility of perspective scientific resources, technologies, and developments—according to the authors of the document.

The process of interaction between state and business should include the leading universities, which train specialists in the sphere of information security. This requires a system of grants for development of perspective means of provision of information security.

Any solution to the problem of tax stimulation of subjects of the digital economy should be built in view of business models. This solution should allow imposing tax on the proper tax base and should be effective in the administrative aspect. At that, there is a possibility of application of separate "tax reaction" to each business model. Analysis of various business models should become a direction of search of approaches to taxation.

There are four such models. In a so called advertising model, which is used by search engines and platforms of social networks, advertising for users brings profit. In this model, deals are performed at the B2B level. In the advertising model, the task of connection of tax to money flow is easily solved.

The things with taxation according to the subscription model scheme are different. In this model, retail consumers are payers. It is considered that in this case tax is more difficult to collect—as compared to a company that uses B2B model.

It is relatively easy to implement taxation for the customer model, in which operations of buy and sell, rent, etc. between users are conducted via a certain digital platform. In a certain sense, this model could be called economy of joint usage. In such business model, the tax could be connected to the flow of revenues from selling the product or service.

The fourth variant is the model of business of online retail seller. The taxation base is easy to establish—it could be sales revenues. The main complexity of this business model is establishment of taxing nexus—i.e., establishment of the fact of presence of business on this territory. Unlike the non-digital equivalent of such business, taxing nexus of a company without its physical presence is not easy to determine.

The given business models should be used during planning of measures aimed at stimulation of development of digital technologies and their usage in various sectors of economy. For example, the Forecast of socio-economic development of the RF for 2017 and the planned period of 2018 and 2019 should contain distribution of usage of information technologies in the socio-economic sphere, state management, and business and should consider the restraining factors, including deficit of personnel, insufficient level of specialists' training, and insufficient number of research of the global level.

5   Conclusions

Generalizing the above measures of tax stimulation, it is possible to group peculiar signs that are formed in the conditions of increase of the share of the digital economy in traditional economy and lead to necessity for reconsidering a lot of approaches in taxation (at this stage, changes influence certain provisions of conventions on exclusion of double taxation and approaches to regulation of transfer pricing):

– Neutrality—systems of taxation should be neutral as to various types of E-commerce—i.e., taxpayers who conduct similar operations have to bear similar tax responsibilities;
– Effectiveness—expenditures of taxpayers for observing the requirements of tax law and administrative costs of tax bodies should be brought down to the minimum;
– certainty and simplicity—the rules should be clear and simple for taxpayers' determining tax liabilities of performed transactions, including time, place, and order of tax calculation;
– justice—rules of taxation should ensure timely calculation of correct sums by taxpayer—i.e., opportunities for tax evasion should be brought down to the minimum;
– flexibility—the taxation system should be flexible and dynamic and at the same level of development in the sphere of technologies.

As a result of implementing the Program "Digital economy of the RF", the following goals will be achieved: creation of the eco-system of the digital economy of the RF, in which the data in the digital form are a key factor of production in all spheres of socio-economic activities and which ensures effective interaction, including transborder, of business, scientific community, state, and citizens; creation of necessary and sufficient conditions of institutional and infrastructural character, elimination of existing obstacles and limitations for creation and development of high-tech businesses and prevention of appearance of new obstacles and limitations in traditional and new spheres and high-tech markets; increase of competitiveness of certain spheres of the Russian economy and the economy on the whole.

Distinguishing specific measures that determine the contents of the mechanism of tax stimulation, it is necessary to note the following:

1. Projects for implementing reduced taxation of revenues from using rights for intellectual activities' results (Patent Box) and specification of parameters of companies' applying increasing coefficient to expenditures for R&D and purchase of rights for results of intellectual activities during calculation of corporate tax.

2. Optimization of application of reduced rate of insurance fees for Internet companies and developers of software, introduction of special tax subsidies for personal income tax, simplification of receipt of Russian tax residency for individuals—highly-qualified specialists, and expansion of deduction of incoming VAT by Russian companies during export of IT and electronic services.

3. Conditions for taxation of Russian online retailers that stimulate development of E-commerce and specification of parameters of taxation with VAT of the services that are provided in the electronic form by foreign sellers.

4. Objects of IT and telecommunication infrastructure are to be included into the list of possible objects of agreements that are envisaged by the law of PPP and the law on concession agreement.

5. The tasks of stimulation include provision of preferences for computer, server, and telecommunication equipment, means of information security of domestic production during conduct of purchases for state and municipal needs, and determination of the approach for interoperability of the program code that is developed within the state order.

6. Reduction of rate for tax on revenues from intellectual property from current 20 to 5%. According to the project's developers, such step will increase the number of registered patents, reduce the outflow of intellectual property abroad, and stimulate foreign companies that possess results of intellectual activities (patents, software) to become tax residents of the RF.

7. Establishment of tax deduction from the personal income tax sum in the volume of 50% of investments into startups and as to subsidies for business angels.

Thus, tax stimulation in the technological sphere and creation of software is a top-priority task that stimulates acceleration of development of the digital economy in Russia. It is important to create legal conditions that improve conditions for business, scientific organizations, and startups that conduct their activities in the information economy.

References

Litvinova, T. N., Morozova, I. A., & Pozdnyakova, U. A. (2019). Criteria of evaluation of effectiveness of industry 4.0 from the position of stimulating the development of knowledge economy. In *Studies in Systems, Decision and Control: Vol. 169. Industry 4.0: Revolution of the 21st century* (pp. 101–111). ISSN 2198-4182. URL: https://doi.org/10.1007/978-3-319-94310-7.

Medvedev, D. A. (2017). Program "Digital economy". Decree dated July 28, 2017, No. 1632-r. Government of the RF, 88 p. URL: http://static.government.ru/media/files/9gFM4FHj4PsB79 I5v7yLVuPgu4bvR7M0.pdf.

Pozdnyakova, U. A., Golikov, V. V., Peters, I. A., & Morozova, I. A. (2019). Genesis of the revolutionary transition to industry 4.0 in the 21st century and overview of previous industrial revolutions. In *Studies in Systems, Decision and Control: Vol. 169. Industry 4.0: Revolution of the 21st century* (pp. 11–21). ISSN 2198-4182. URL: https://doi.org/10.1007/978-3-319-94310-7.

Taxation of the Internet Enterprise as Key Subjects of Information Economy

Tatyana N. Aksenova, Kita K. Bolaev, Elena V. Samaeva
and Burgsta E. Evieva

Abstract *Purpose* The purpose of the work is identifying the role and importance of Internet enterprises for the formation and development of the information economy in modern Russia and to develop recommendations for improving their taxation system in order to help build the Russian information economy. *Methodology* The methodology of this research is based on the application of economic statistics methods to analyze the role and importance of Internet enterprises for the information economy of modern Russia. At the same time, the information materials of the World Economic Forum, the Business and Technology Information Agency (VC), and the Federal State Statistics Service (Rosstat) for 2018 are the information and analytical base. *Results* In the course of the project it is proved that in the modern Russian information economy, Internet entrepreneurship plays an important role, acting as a key business entity, replacing traditional business, and is crucial for its construction. *Recommendations* In order to activate Russian Internet entrepreneurship and increase its global competitiveness, it is recommended that a special tax regime be established that automatically notifies the tax authorities of payment transactions and automatically collects profit tax at a reduced rate (3%). To collect tax payments from cross-border trade on the Internet, it is also recommended to create a special tax regime for automated taxation of foreign Internet enterprises at a rate of 0.5% of income. The logic and mechanism for the action of the author's recommendations reflects the model presented by the taxation of Internet enterprises in order to help build the Russian information economy.

T. N. Aksenova (✉) · K. K. Bolaev · E. V. Samaeva · B. E. Evieva
B.B. Gorodovikov Kalmyk State University, Elista, Russian Federation
e-mail: tn.aksenova@gmail.com

K. K. Bolaev
e-mail: bolaev_kk@mail.ru

E. V. Samaeva
e-mail: samaeva@mail.ru

B. E. Evieva
e-mail: evievaburgsta@yandex.ru

© Springer Nature Switzerland AG 2019   199
I. V. Gashenko et al. (eds.), *Optimization of the Taxation System: Preconditions, Tendencies, and Perspectives*, Studies in Systems, Decision and Control 182, https://doi.org/10.1007/978-3-030-01514-5_23

Keywords Taxation · Internet enterprises · Information economy
Modern Russia

JEL Classification E62 · H20 · K34

1 Introduction

Under the conditions of the information economy there are intensive transforma-
tional processes in the system of entrepreneurship, an important place among which
is the formation and development of Internet entrepreneurship. It is conditioned,
first, by the desire of entrepreneurs themselves to optimize economic activity.
Compared with the traditional Internet enterprises can minimize the stock of production and, accordingly, the risks of sales, as well as production costs due to the
minimum requirements for fixed and working capital. Moreover—the Internet
enterprise can successfully function without employees, which eliminates the cost
of labor. At the same time, the geography of business is expanding and the sales
volume can significantly increase.

Secondly, under the conditions of the information economy, there is a change in
consumer preferences and an increase in demand for the products of Internet
enterprises. Consumers value their convenience and comfort highly, preferring to
choose the necessary products through a highly efficient automated search system
that allows them to study all available offers and choose the best option for satisfying their needs. At the same time, the production of Internet enterprises, as a rule,
has price advantages with a similar quality and is therefore more preferable for
consumers.

Thirdly, under the pressure of competition from existing and increasing in the
number of Internet enterprises, more and more traditional enterprises are compelled
to supplement or transfer their business to the Internet. In the context of the
information economy, there are great opportunities for Internet entrepreneurship
(due to the availability of software and software) and the purchase of products on
the Internet (due to the mass availability of high-speed Internet). Therefore, more
and more enterprises prefer to conduct business on the Internet (at least in addition
to the main traditional business), displacing completely traditional businesses from
the market.

A key aspect of the functioning and development of Internet enterprises is
taxation, since it defines the institutional environment for starting and running
businesses of these enterprises. The working hypothesis of this study is the
assumption that in modern Russia the absence of special taxation conditions is a
deterrent to the development of Internet entrepreneurship, which in turn hinders the
construction of an information economy. The purpose of the work is to identify
the role and importance of Internet enterprises for the formation and development of

the information economy in modern Russia and to develop recommendations for improving their taxation system in order to help build the Russian information economy.

2 Materials and Method

The object of this study is Internet entrepreneurship, which is understood as the conduct of business on the Internet on the following conditions:

- lack of own production and specialization in the trade of finished products purchased from suppliers on the basis of "just-in-time" (lack of stocks, and ordering products from suppliers as orders are received from consumers);
- the absence of any element of traditional entrepreneurship and the maintenance of fully Internet entrepreneurship (lack of fixed assets and a minimal set of enterprise assets limited to a computer, Internet access and software).

As you can see, in the category of Internet entrepreneurship, traditional businesses that have an Internet site and conduct Internet commerce in parallel with the main traditional business do not enter this work. This is due to the fact that, in the taxation aspect, these enterprises should be referred to traditional business, and they are common in the post-industrial economy, while fully Internet enterprises that do not assume the traditional business are characteristic of the information economy and potentially need special conditions taxation due to the specifics of its activities (lack of fixed assets, maximum simplicity of business processes).

The peculiarities of making business by Internet enterprises, as well as the trend of development of Internet entrepreneurship in modern economic systems, are discussed in the works of scientists such as Chen and Ku (2016), Ji et al. (2018), Shao et al. (2018), Xiang et al. (2017), Yu et al. (2017). The importance of Internet entrepreneurship as a manifestation and factor in the formation and development of the information economy is emphasized in the writings of such scientists as Bogoviz et al. (2019), Lobova et al. (2019), Ragulina et al. (2019), Sukhodolov et al. (2018a, b, c).

At the same time in existing studies and publications, insufficient attention is paid to the taxation of the Internet enterprise. To fill this gap our work is devoted. The methodology of this research is based on the application of economic statistics methods to analyze the role and significance of Internet enterprises for the information economy of modern Russia. At the same time, the information materials of the World Economic Forum, the Business and Technology Information Agency (VC), and the Federal State Statistics Service (Rosstat) for 2018 are the information and analytical base.

3 Results

Our review and content analysis of the available statistical data made it possible to obtain the following results:

- According to Rosstat, in 2018 (as of the end of 2017), the number of wholesale and retail trade enterprises in Russia totaled 1465.1 thousand units, that is 32% of the total number of enterprises (4,561,000). Turnover of trade enterprises in 2018 (RUB 57,830.4 billion) increased by RUB 24.47 trillion, (73%) as compared to 2010 (RUB 33,359.8 billion) (Rosstat 2018);
- Turnover of Internet trade in Russia in 2017 amounted to RUB 1040 billion. Compared to 2010 (260 million rubles). It increased 4-fold. At the same time, the share of cross-border trade transactions increased from 8% in 2010 to 36% in 2017. The majority (91%) of cross-border trade transactions were made by Russian buyers with Chinese Internet enterprises (Information Agency "Business and Technology" (VC) 2018).
- According to the Global Information Technology Report prepared in the framework of the World Economic Forum in 2016, in Russia, among indicators of readiness for building the information economy, the lowest value was assigned to the indicator of using the latest information and communication technologies in business (7th pillar: Business usage)—3, 6 points out of 7 possible (67th place among 139 countries of the world). In terms of this indicator, Russia also lags behind the countries from the High-income group average (4.9 points) (World Economic Forum 2018).

It allowed us to conclude that Internet entrepreneurship is dynamically developing in modern Russia, but so far its share is rather small and amounts to 2% of the total turnover of trade enterprises. The low level of development of Internet entrepreneurship is a barrier to building the information economy. Preference by Russian consumers to cross-border trade operations (with a longer delivery time and lack of guarantees of consumer rights under Russian law) attests to the low global competitiveness of Russian Internet entrepreneurship.

Currently, Internet enterprises in Russia are not allocated to a separate category of taxpayers and fall either under the general taxation regime, or under special taxation regimes (subject to registration as subjects of small and medium-sized businesses). It causes the following problems:

- High level of taxation of Russian Internet enterprises: in addition to the profit tax (or its analogues), Internet enterprises pay value-added tax, personal income tax and social contributions;
- The high complexity of taxation of Russian Internet enterprises: the need to maintain tax reporting under general rules does not allow fully automating the activities of Internet enterprises (which is their main competitive advantage compared to traditional business);
- Low competitiveness of Russian Internet enterprises: fully and on a general basis, following the current Russian legislation on consumer rights (including

return of goods, guarantee of its quality, etc.), and also paying a full set of taxes, Russian Internet enterprises give way to foreign market positions Competitors with significant price advantages (since Russian taxes do not pay);

- Insecurity of the rights of the majority of Russian buyers of products of Internet enterprises: when purchasing products from foreign enterprises, Russian consumers bear significant risks associated with the violation of the delivery terms of goods, their inconsistency with the declared quality, and the difficulty or even impossibility of returning goods and obtaining guarantees for them;
- Lack of receipt of tax revenues by the state budget of the Russian Federation: an unprofitable tax regime hinders the development of domestic Internet entrepreneurship, while cross-border trade transactions on the Internet are not subject to Russian taxes. At the same time, the complexity of tax administration of Internet enterprises causes the risk of their future shadowing.

To solve these problems in order to facilitate the construction of the Russian information economy, we developed the following model for taxation of Internet enterprises (Fig. 1).

Figure 1 shows that improving the taxation system of Internet enterprises in order to help build the Russian information economy, we propose the following recommendations:

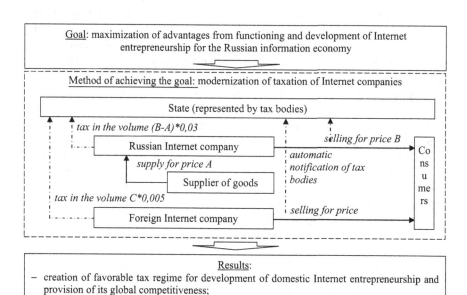

Fig. 1 The taxation model of Internet enterprises in order to promote the construction of the Russian information economy. *Source* Compiled by the authors

- the introduction of a special tax regime for Russian Internet enterprises, within which no value added tax is paid, and only the income tax at the rate of 3% is paid;
- introduction of a special tax regime for foreign Internet enterprises, within which no value added tax is paid, and only a tax on income at a rate of 0.5% is paid;
- automatic notification of tax authorities about payment transactions made by Internet enterprises and automatic collection of taxes without the need to maintain and provide additional tax reporting.

This allows to maximize the benefits of the functioning and development of Internet entrepreneurship for the Russian information economy, by creating a favorable tax regime for the development of domestic Internet entrepreneurship and ensuring its global competitiveness, as well as increasing tax revenues to the state budget, both from Russian and foreign Internet enterprises.

4 Conclusion

Thus, in the course of the research, the hypothesis put forward is confirmed and proved that in the modern Russian information economy, Internet entrepreneurship plays an important role, acting as a key business entity, replacing traditional business, and is crucial for its construction. To activate Russian Internet entrepreneurship and increase its global competitiveness, it is recommended to create a special tax regime that automatically informs the tax authorities about the payment transactions to be performed and automatically levies a profit tax at a reduced rate (3%).

To collect tax payments from cross-border trade on the Internet, it is also recommended to create a special tax regime for automated taxation of foreign Internet enterprises at a rate of 0.5% of income. The logic and mechanism for the action of the author's recommendations reflects the model presented by the taxation of Internet enterprises in order to help build the Russian information economy. Additional revenues collected in the state budget can be directed to the development of the information economy infrastructure.

References

Bogoviz, A. V., Lobova, S. V., Ragulina, Y. V., Chernitsova, K. A., & Shkodinsky, S. V. (2019). Internet tools for development of knowledge economy: Essence, tendencies, and perspectives. *Advances in Intelligent Systems and Computing, 726*, 39–45.

Chen, T.-J., & Ku, Y.-H. (2016). Rent seeking and entrepreneurship: Internet startups in China. *Cato Journal, 36*(3), 659–688.

Federal State Statistics Service. (2018). Russia in numbers: A short statistical collection. http://www.gks.ru/wps/wcm/connect/rosstat_main/rosstat/ru/statistics/publications/catalog/doc_1135075100641. Data accessed: 09.07.2018.

Information Agency "Business and Technologies" (VC). (2018). Market of internet trade in Russia in 2017. https://vc.ru/35781-rynok-internet-torgovli-v-rossii-v-2017-godu-prevysil-1-trln-rubley. Data accessed: 09.07.2018.

Ji, Y., Jiang, Y., & He, L. (2018). An evaluation method based on co-word clustering analysis—Case study of internet + innovation and entrepreneurship economy. *Communications in Computer and Information Science, 849,* 588–595.

Lobova, S. V., Ragulina, Y. V., Averin, A. V., Simonov, S. Y., & Semenova, E. I. (2019). Methods of digitization of the Russian economy with the help of new internet technologies. *Advances in Intelligent Systems and Computing, 726,* 221–228.

Ragulina, Y. V., Semenova, E. I., Avkopashvili, P. T., Dmitrieva, E. A., & Cherepukhin, T. Y. (2019). Top-priority directions of implementing new internet technologies on the territories of rapid economic development. *Advances in Intelligent Systems and Computing, 726,* 182–188.

Shao, Y., Wu, T., Qiu, H., & Wang, Z. (2018). Ambidextrous activities of internet-based entrepreneurships in Apple App Store: Two sides of user feedback. *Technology Analysis & Strategic Management, 2*(1), 1–16.

Sukhodolov, A. P., Popkova, E. G., & Kuzlaeva, I. M. (2018a). Methodological aspects of study of internet economy. *Studies in Computational Intelligence, 714,* 53–61.

Sukhodolov, A. P., Popkova, E. G., & Kuzlaeva, I. M. (2018b). Peculiarities of formation and development of internet economy in Russia. *Studies in Computational Intelligence, 714,* 63–70.

Sukhodolov, A. P., Popkova, E. G., & Kuzlaeva, I. M. (2018c). Production and economic relations on the internet: Another level of development of economic science. *Studies in Computational Intelligence, 714.*

World Economic Forum. (2018). The global information technology report 2016 innovating in the digital economy. http://www3.weforum.org/docs/GITR2016/WEF_GITR_Full_Report.pdf. Data accessed: 09.07.2018.

Xiang, Y., Chen, X., Mei, L., & Chen, J. (2017, January). Influence of social networks, opportunity identification on the performance of internet entrepreneurship: The evidence of Zhejiang province in China. In *PICMET 2017—Portland International Conference on Management of Engineering and Technology: Technology Management for the Interconnected World, Proceedings* (pp. 1–8).

Yu, X., Roy, S. K., Quazi, A., Nguyen, B., & Han, Y. (2017). Internet entrepreneurship and "the sharing of information" in an Internet-of-Things context: The role of interactivity, stickiness, e-satisfaction and word-of-mouth in online SMEs' websites. *Internet Research, 27*(1), 74–96.

The Model of Well-Balanced Taxation for Overcoming the Shadow Economy in Modern Russia

Ulyana A. Pozdnyakova, Aleksei V. Bogoviz, Svetlana V. Lobova,
Julia V. Ragulina and Elena V. Popova

Abstract The chapter dwells on theoretical foundations of the shadow economy, considers the types of existence and growth of the shadow sector, evaluates the scale and role of the shadow economy, and studies pros and cons of existence of the shadow economy, as well as peculiarities of the model rows of measuring the shadow economy, related to tax evasion at the national level. For adapting these methods to the regional level, correspondence of the national methods and indicators in the regional aspect is viewed, and the most adequate indicators are selected, which are presented in the form of the model of well-balanced taxation. Practical implementation of the model will allow regulating the volumes of the shadow economy for making more justified decisions on improvement of the tax policy at various levels of management.

Keywords Taxes · Taxation · Shadow economy · Russia

U. A. Pozdnyakova (✉)
Volgograd State Technical University, Volgograd, Russia
e-mail: ulyana.pozdnyakova@gmail.com

A. V. Bogoviz · J. V. Ragulina
Federal State Budgetary Scientific Institution "Federal Research Center
of Agrarian Economy and Social Development of Rural Areas—All Russian
Research Institute of Agricultural Economics", Moscow, Russia
e-mail: aleksei.bogoviz@gmail.com

J. V. Ragulina
e-mail: julra@list.ru

S. V. Lobova
Altai State University, Barnaul, Russia
e-mail: barnaulhome@mail.ru

E. V. Popova
Plekhanov Russian University of Economics, Moscow, Russia
e-mail: epo495@gmail.com

© Springer Nature Switzerland AG 2019
I. V. Gashenko et al. (eds.), *Optimization of the Taxation System: Preconditions,
Tendencies, and Perspectives*, Studies in Systems, Decision and Control 182,
https://doi.org/10.1007/978-3-030-01514-5_24

1 Introduction

Shadow business envisages primarily tax evasion. As practice shows, tax evasion in the most peculiar sign of the shadow economy. Evasion from other payments takes place according to the similar mechanism, but is rarer. On the contrary, tax evasion is very popular.

The current economic situation in Russia leads to the fact that a lot of companies—regardless of their purpose, volumes of issued products, and sphere of activities—use tax evasion purposefully. This is due to large tax rates, which make entrepreneurs hide some of their revenues for continuation of functioning of their businesses.

The part of economy that includes initiative economic relations between citizens or their associations is real activities of people for provision of various services for population; it cannot be treated as negative and undesirable for society. However, if such legal activities are connected to gaining the income that is not controlled by the state and tax evasion, the relations that are envisaged by the Criminal Code appear. It's not that entrepreneurial activities as such create an offence but unfair entrepreneurship and violation of established rules of conducting economic activities. Depending on the level of public danger of this violation, administrative or criminal responsibility is assigned, and the conducted damage is compensated. There's another attitude towards the part of the shadow economy that uses human vices and serves destructive public needs that can morally and physically deform a person. The corresponding activities here are socially dangerous and criminal. Thus, the structure of the modern Russian shadow economy could be presented in the form of three elements:

1. "Underground" economy—entrepreneurial and other economic activities, oriented at satisfaction of normal public needs; however it is conducted in the forms that are uncontrolled by the state for the purpose of gaining uncontrolled income and tax evasion.
2. "Fictitious" economy—entrepreneurial and other economic activities, related to various forms of fraud and con game.
3. Criminal economy that is oriented at satisfaction of destructive public needs.

The main reasons of existence and development of the shadow economy are instability and imbalance of official economy, which is in a deep crisis, and ineffectiveness of tax policy of the state. A certain part of entrepreneurs are ready to legalize their business—but only under the condition of reduction of tax load to the acceptable level.

One of the negative features of regional shadow schemes is that they lead to deficit of local budgets, as at the modern stage the shadow economy is related primarily to non-payment of taxes into budgets of various levels. At the same time, tax evasion not only hinders the execution of budgets but also distorts the system of taxation and its executing the fiscal and regulating functions, including at the local level. All this slows down the development and improvement of local taxation. One of the first steps for fighting this problem is adequate evaluation of the scale and structure of the shadow economy of the region in view of the taxation factor (Allingham 1972).

2 Materials and Methods

Theoretical and empirical tools of managing the shadow economy at the national level are already created; they are described in the works of V. P. Vishnevsky, Y. B. Ivanova, S. N. Kovaleva, A. V. Kostina, Y. V. Latova, I. A. Mayburova, V. V. Popova, N. I. Suslova, S. Davids, C. Elgin, E. L. Feige, B. S. Frey and H. Week, P. M. Gutmann, F. Schneider, D. Kaufmann and A. Kaliberda, R. Klinglmair, M. Lacko, V. Tanzi, et al. However, the methods of measuring and managing the shadow economy at the macro-level are not always effective in the regional aspect, which is caused by three reasons: firstly, the character of this phenomenon, which envisages tax evasion; secondly, specific peculiarities of the shadow economy of each region; thirdly, problems with statistical data that are necessary for calculation.

That's why the issue of development of approaches to managing the main quantitative parameters of the shadow economy at the regional level with the usage of open sources of information requires to be studied.

At present, three groups of methods are used for evaluation of the volumes of the shadow economy at the national level:

1. Direct—micro-economic methods, based on voluntary surveys or usage of the data of tax audit.
2. Indirect (macro-economic) methods—economic and other indicators containing the information on development of the shadow economy in time.
3. Model methods. There are two variants of the model approach—Multiple Indicator-Multiple Cause (MIMIC) and Dynamic Multiple Indicator-Multiple Cause (DYMIMIC).

These approaches use structural and econometric models in which the share of the shadow sector is an indirectly assessed variable, which allows considering all main factors that influence the development of the shadow economy.

Analysis showed that advantages of the model approach during determining the size of region's shadow economy excess drawback of its usage. The main advantage of the methods of modeling for evaluation of region's shadow economy consists in varying set of variables and indicators. For modeling at the level of regions, which differ by econometric characteristics, level of accessibility of data, and the studied period, the latter is the key factor (Giles and Tedds 2002).

It is supposed that the size of the shadow economy of region could be determined indirectly with the help of aggregate indicators. This allows determining structural dependencies of the size of the shadow economy on casual variables, which could be convenient for forecasting its sizes in the future.

Change of the values of causal variables leads to changes of the value of the hidden variable (size of region's shadow economy). Analysis of general variation of causal variables and indicators shows the change of the hidden variable; this allows concluding on dynamics and relative scales of the region's shadow economy.

In the DYMIMIC models, three types of causal variables are distinguished:

1. The load of direct and indirect taxation, the load of nominal and real taxation. It is supposed that increase of the level of taxes is a significant reason for transfer to the shadow sector.
2. The load of administrative regulation. It is considered that increase of the load of regulation from public authorities is also a reason for transfer into the shadow sector.
3. Tax moral—citizens' attitude towards the state; it describes the individuals' readiness (at least, potentially) to change activities in the official sector to shadow activities. Accordingly, it is considered that reduction of the level of tax moral leads to increase of the size of the shadow economy.

The indicators of DYMIMIC models include the following:

1. Monetary indicator. It is supposed that growth of the shadow economy leads to growth of the volumes of monetary operations—especially if cash is used for preventing the detection.
2. Indicator of labor market. Reduction of the level of economically active population, employed in the official economy, leads to growth of this indicator in the shadow economy. Thus, increased activity in the shadow sector is reflected in reduction of labor expenses and labor time in the official economy.
3. Indicator of production. Growth of the size of the shadow economy leads to outflow of production factors (resources) from the official economy, which will negatively influence the growth rates of the latter (Tanzi 1999)

In view of analysis of studies of the shadow economy at the national level with the usage of the MYMIC and DYMIMIC models, their indicators were adapted to the regional level in view of the taxation factor.

The shadow economy ousts formal mechanisms of taxation. Shadow activities distorts the production process in the legal economy, weakens labor motivation of employees, and leads to loss of their qualification (Tanzi 1999).

In the Soviet economic system, the shadow economy performed two main functions—economic and public. The shadow economy ensures a public niche for active people who were not able to realize themselves in the official structures.

In the conditions of modern Russia, the shadow economy is the main tool of supporting economic and public balance, creating conditions for survival of business and population. As of now, shadow factors are immanent for economic ties and relations. The very existence of the legal economic system in its current form is impossible without the shadow world.

Constructive aspects in the shadow economy were noted by Western scholars. Thus, the Nobel Prize winner, Vasily Leontyev, noted that the shadow economy is not a reason for panicking. It would invest capital into the open sector, creating new jobs and improving infrastructure. From the point of view of economic subjects, the shadow economy allows working rationally, as it saves money on taxes. The shadow economy also expands the probability of additional earnings for employees of officially registered companies. According to the Swedish economist A. Aslund, the shadow economy in Russia—unlike the existing Western stereotypes—produces legal services and goods. Criminal groups account only for 3% of the shadow sector. Thus, creating new jobs and sources of income, the shadow economy performs the role of public stabilizer, levels inequality of incomes, and reduces public tension on society (Dell'Anno 2007).

Depending on the motivation systems, the following main forms of shadow economic activities are distinguished:

1. Pseudo-shadow or half-shadow activities (design of shadows). Taxes reduce income—therefore, they should be avoided. The example of such behavior of economic subjects is mass registration of small companies at the initial period of announced tax subsidies and their mass closure after the ends of "tax vacations". This action is absolutely legal—as it does not avoid the requirements of the tax system but adapts to it. It acquires the half-shadow form, as it could be accompanied by "adaptation" of indicators of activities to the required ones.
2. Forced (rational) shadow activities. As the price of lawful behavior exceeds the corresponding price in the conditions of shadow activities, this type of shadow activities is not desirable, has a forced motive, and is provoked by inadequate tax policy of the state. It is considered that this type of shadow activities could transfer to transparent economy with the corresponding conditions, which is a basis for offers on legalization of the shadow sector and its integration with the legal sector.
3. Irrational shadow activities are the result of irrational behavior, which subjects is inclined to risk and risky type of behavior. This position is supported by the theory of economic crime, which is based on the rule of solving of economic crimes and inevitability of sanctions. According to this theory, shadow economic activities are unavoidable and do not have motivation—there's only result of self-realization of a specific public and psychological type of subject of economic activities.

All the above forms of shadow activities envisage conscious selection, justified or unjustified, correct or incorrect, voluntary or forced (Buehn et al. 2009). In all three cases, legal transparent action is allowable, but difficult.

The main means of tax evasion, regardless of the tax type, are as follows:

Hiding the taxation objects: distortion or non-reflection of financial and economic operations in financial accounting; conduct of financial and economic activities without necessary registration in the State tax inspection or license, and usage of fake documents and organizations; conduct of financial and economic activities via accounts of other organizations or structural departments; destruction of financial documents after the operation.

Reduction of taxation objects: transfer of a part of money to different financial accounts; creation of fictitious excess of products by increasing the losses, unjustified writing off of final products; providing distorted data in the financial documents.

The institute of tax evasion allows hiding a part of financial flows and property in the shadow, balancing at the edge of legal and illegal actions (Giles and Tedds 2002). Thus, with fictitious operations, financial documents reflect non-existing movement of products and services between the company or intermediary. For example, money is transferred for the products that were not supplied, fictitious marketing services are ordered, and fake exports is conducted, which allows for VAT return.

A lot of managers of companies use the status of individual entrepreneur without registration of a legal entity for large reduction of paid taxes due to subsidies that are envisaged by the Federal law for subjects of small entrepreneurship. Another example of tax evasion with fictitious operations on the edge of legal and illegal actions could be the situation when a travel company purchases several buses but registers them as private property of the company's managers' relatives, who, in their turn, officially allow hired drivers to use the buses and conclude rental agreements for the buses. As a result, the buses are not reflects in the company's balance sheet, and taxes are not paid. It should be noted that a lot of violations are observed not with the optimization schemes but with accompanying conditions of this scheme. For example, usage of subsidies, provided to small companies, is often accompanied by falsification of the number of the organization's employees. This automatically makes the scheme illegal.

Besides, non-equivalent exchanges are used, during which unprofitable contracts in favor of third parties are concluded, and cost proportions are purposefully distorted. For that, barter exchanges and different systems are used, which allow purchasing and selling products and services for the prices that are lower or higher than their market price: systems of reduced transfer prices are used; debt securities are issued with further play with differences between their nominal and market price; shares of different liquidity level are exchanged; loans with increased interest are issued.

Most of the above schemes are based on the following main principles: separation of companies that produce added value from selling their products and from financial flows by creation of different structures; division of assets and liabilities of manufacturing companies with further transfer of assets to intermediary structures and putting liabilities onto the state and creditors via the practice of non-payment; usage of official accounting at manufacturing companies and double accounting in intermediary structures with further destruction of the part of the documents.

Outlines of managerial structures could have closed and open form. Within the closed contour, money flows pass along the circle and return to the initial company that produces added value. In this case, money that was hidden from taxation transforms into investments. According to the experts, a large share of foreign investments into the Russian economy accounts for domestic resources that were previously transferred abroad and converted in this way. Within the open contour, resources are not returned to manufacturing companies. They are hidden in accounts of intermediaries and taken abroad or invested into another business, or serve the purposes of personal enrichment of owners. In any model, tax evasion envisages creation of long chains of companies that complicate the possibility of tracking the financial flows. At that, each operation is formally legal, but the scheme is not.

The process of tax evasion usually has a range of stages. In a certain sense, "life cycle" of tax violation is similar to "life cycle" of a product. At the first stage, taxpayer creates a new scheme of tax evasion. At the second stage, this scheme becomes popular among taxpayers. When it becomes known to tax bodies, the third stage starts, in the course of which a way to fight this scheme is searched for, and forms and methods of most effective determination of tax violation are defined. The fourth stage is active and massive opposition of public authorities to violations, which leads to reduction of their number and conducted damage. At the fifth stage, balance between violators and public authorities is set at the least possible level.

3   Results

Well-balanced formation and implementation of the taxation model in the conditions of the shadow economy determines the value of the tax potential. The model of well-balanced taxation is a complex of three interconnected financial and budget spheres of activities, regulated by special legal norms of the whole arsenal of the legal acts of the state. This model should determine establishment and evaluation of planned, factually executed, and forecasted tax liabilities of subjects of tax legal relations (tax planning), conduct of scientifically justified measures of current interference with the course of execution of the country's budgets of stimulating character (tax regulation) and sanction measures in case of violation of the norms of tax law (tax control). Thus, the elements of the model of well-balanced taxation include planning, regulation, and control.

Wise regulation of tax potential includes creation of general tax climate for internal and external activities of organizations and provision of most preferable tax conditions for stimulation of top-priority sectorial and regional directions of capital movement (Johnson et al. 1998).

Tax regulation is conducted in the following ways and methods: ways—subsidies and sanctions, methods—investment tax credit, delay, transfers, etc.

Essential approaches to regulation of tax legal relations in a lot of countries are determined by the Tax Code. The final goal of tax regulation is to balance the interests of three subjects: state, economic subjects, and citizens.

For the purpose of observing the budget's balance, which is set in the budget law, the minimum volume of tax load for citizens and companies should be limited by the sum of state expenditures. Otherwise, growth of tax rates and number of taxes might urge taxpayers to go to the shadow economy, as they will be unable to pay the announced sums. Thus, the planned sum of taxes will not be paid to the budget. Human factor has become very topical. In this case, we speak of the problem of non-payment of taxes, which negatively influences the expansion of region's tax potential. Presence of the shadow sector also influences the formation of tax potential. Finding the optimal ways and methods of fighting the shadow economy allows increasing the tax potential.

Regulation of the tax system of the RF at the modern stage is based on the Tax Code of the RF, according to which the Russian laws on taxes include: the Tax Code and accompanying federal laws on taxes, laws on taxes that are adopted by legislative public authorities of the subjects of the RF, and normative legal acts of representative bodies of local administration.

The main reserves of growth of incomes consist in reduction of debt, improvement of tax administration, increase of fight against legal violations, and increase of sanctions for their execution. If the necessity for covering the growing state expenditures is considered, it is important to pay special attention to the volume of tax revenues, oriented at the maximum possible level of collection of taxes. The indicator that characterizes this value is tax potential. It is necessary to have a certain set of criteria to which the analytical system that is used for characterizing the region's tax potential should conform: dynamics, structure, and territorial differentiation of tax revenues of region and tax bases; factor space of tax potential; quality of region's tax potential.

Besides, tax control has a new sphere—tax monitoring. It is an electronic information interaction. Before filing the tax declaration, a taxpayer can solve the issues of taxation about which he has certain doubts. Also, the tax body in real time can gain access to the data of financial and tax accounting of the taxpayer for verification of correctness and timeliness of reflection of economic operations by the taxpayer for the purposes of taxation.

Market economy has to have succession and stage-by-stage procedures. The issues of taxation have to have specific tasks. Tax potential is different as to the levels of budget and tax systems, spheres of economy, and types of economic activities, types of taxes, economic subjects, groups of taxes, and the volume of factual and optimal tax potential.

Complex characteristics of tax potential is reflected in tax passport. It is one of the main tools that allow taking analysis and forecasting of tax revenues to a new level. Tax passport of the subject of the RF, federal district in Russia, and Russia on the whole contains such data as the volume of industrial products, investments into fixed capital, inflation level, volume of agricultural products, retail product turnover, indicators of financial and economic activities of companies, etc.

Comparison of the practice of taxation in Russia and Western countries shows that depending on expedience, application of the same tools of tax regulation in different countries could have positive or negative effect.

The tools that are used in developed countries include: changes in the structure of the tax system (federal, regional, and local taxes and regulating taxes), changes of tax rates and tax subsidies, creation of free tax zones, offshore zones, and scientific towns.

For determining the efficiency of tax regulation, it is necessary to determine the level of its correspondence to the criteria of sufficiency and effectiveness. This indicator determines two optimal admissible limits of taxation: upper and lower. The lower limit is the level of budget needs. In this context, taxation cannot be lower than the threshold of budget needs for financing of the main functions of the state. The upper limit envisages the level of payment capacity of taxpayers. Taxation cannot exceed the allowable level due to the risk of reduction of taxpayer's incomes and, therefore, tax revenues—and this may lead to bankruptcy of the taxpayer and termination of inflow of tax revenues.

Regulation is an important tool in the process of formation of state revenues. The necessity of control over the paid sum and potential sum of revenues into the budget stimulates the reduction of difference between these indicators. Regulation forms responsibility with taxpayers, and is one of the three main functions that are parts of the tax mechanism.

The research shows that the shadow economy stimulates the increase of well-being population only in case of single representative of population, and the shadow economy decreases population's well-being if the population is represented by a lot of agents, and the volume of the shadow economy depends on the number of population in the country, with all other conditions being equal. For reduction of the volume of the shadow economy, it is necessary to reduce tax load and improve the quality of state services. Reduction of the volume of the shadow economy will be stimulated by expansion of authorities of regions for collection and usage of collected taxes. Decentralization of taxpayers' groups transforms the relations state-population as to relation region-population—i.e., to reduction of the number of agents, which, within the model's hypotheses, leads to reduction of the share of the shadow economy, with all other conditions being equal.

References

Allingham, M., & Sandmo, A. (1972). Income tax evasion: A theoretical analysis. *Journal of Public Economics, 1,* 323–338.

Buehn, A., Karamann, A., & Schneider, F. (2009). Shadow economy and do it yourself activities: The German case. *Journal of Institutional and Theoretical Economics JITE, 165*(4), 701–722.

Dell'Anno, R. (2007). The shadow economy in Portugal: An analysis with the MIMIC approach. *Journal of Applied Economics, X*(2), 253–277.

Giles, D., & Tedds, L. (2002). Taxes and the Canadian Underground Economy. In *Canadian tax paper* (Vol. 106, 270 p.).

Johnson, S., Kaufmann, D., & Zoido-Lobaton, P. (1998). Regulatory discretion and the unofficial economy. *American Economic Review, 88*(2), 387–392.

Tanzi, V. (1999). Uses and abuses of estimates of the underground economy. *The Economic Journal, 109,* 338–347.

The Strategy of Provision of Tax Security of the State in the Conditions of Information Economy

Gilyan V. Fedotova, Ruslan H. Ilyasov, Anastasia A. Gontar
and Viktoria M. Ksenda

Abstract *Purpose* The purpose of the chapter is to analyze the current processes of informatization of the Russia's tax system, to study the development and implementation of new digital services for taxpayers, and to evaluate effectiveness of collection of taxes due to introduction of additional remote technologies into the practice of taxation. Tax policy of the state is closely interconnected to replenishment of the state budget, which is a guarantor of state's execution of its obligations. Efficiency of tax policy influences the financial state of the country. That's why building a strategy of increase of tax security is an important task of the national scale. *Methodology* The research is performed with the help of graphical presentation of information, trend analysis, comparison, analogy, and systematization. *Results* According to the peculiarities of the built model of the information economy, we consider the main directions of development of the existing tax policy, conduct evaluation of threats to the level of collection of taxes and fees in the country, and study the notion and indicators of the quality of taxation in Russia and certain countries of the world. *Recommendations* The results of analysis of the existing system of taxation could be used in the system of strategic planning of socio-economic development of Russia for the future planned periods. Besides, the received recommendations for development and formation of the strategy of provision of tax security allow for further correction of the system of economic security of the state, and the applied methods and means of analysis of the level of collection

G. V. Fedotova (✉) · A. A. Gontar
Volgograd State Technical University, Volgograd, Russian Federation
e-mail: g_evgeeva@mail.ru

A. A. Gontar
e-mail: 261984@mail.ru

R. H. Ilyasov
Chechen State University, Grozny, Russian Federation
e-mail: ilyasov_95@mail.ru

V. M. Ksenda
Volgograd State University, Volgograd, Russian Federation
e-mail: KsendaVM@volsu.ru

© Springer Nature Switzerland AG 2019
I. V. Gashenko et al. (eds.), *Optimization of the Taxation System: Preconditions, Tendencies, and Perspectives*, Studies in Systems, Decision and Control 182, https://doi.org/10.1007/978-3-030-01514-5_25

of taxes could be implemented into the practice of work of tax bodies within implementation of the information economy model.

Keywords Taxes · Tax system · Threats · Economic security
Tax risks

JEL Classification H21 · H22 · H26 · H30 · H56

1 Introduction

The current processes of informatization and digitization of all spheres of national economy require new approaches and tools for subjects' performing their current activities. All growing threats from the information influence on existing systems and risks, related to possible losses, require from state bodies higher technological advantage. The course at increase and provision of economic security of the state is a complete complex work of all government bodies for preservation of single economic space within the country, maximum protection of its subjects, reduction of unfair competitive struggle, and opposition to implemented economic sanctions against Russia. In this struggle, large role belongs to state support for all implemented measures—primarily, financial. That's why formation of effective, rational, efficient, and high-quality tax system within the country will allow solving a complex of problems related to insufficient collectability of taxes and fees, tax evasion of economic subjects, ineffective and unjustified increase of tax load, and reduction of bureaucratic influence in relations "tax body–taxpayer".

Application of new information achievements will increase the level of tax security in the country and will allow Russia to successfully compete in the international markets of resources and final products. The final goal is increase of population's living standards on the basis of using and implementing information and communication technologies. For this, it is necessary to pay attention to official documents of strategic character, which are developed and approved at the state level and which regulate target landmarks for future information economy. These documents include the Strategy of development of information society in the Russian Federation for 2017–2030, for implementation of which the National Programs (hereinafter—NP) are implemented. Let us pay attention to the NP "Information society" for 2011–2020, which already has certain results.

2 Materials and Method

Theoretical and applied issues of evaluation of effectiveness of state management and budget and tax policy are viewed in the works of a lot of Russian and foreign authors—e.g., Aristovnik and Obadić (2015), Bondarenko (2015), Comerio and

Batini (2016), Vartakova et al. (2016), Plotnikov et al. (2015)), Méndez Reátegui et al. (2016), Pessoa et al. (2016), and Romanova et al. (2017). Scientific and methodological issues of managing the process of informatization of socio-economic systems are studied in the works: Dai (2013), Hawes and Li (2017), Huggins and Frosina (2017), Nasir et al. (2017), Sakil (2017), Sukhodolov et al. (2018).

However, despite the large number of publications on adjacent topics, the issues of provision of tax security and its increase in the modern conditions of implementation of the information model of economy and the role of tax administration in this model are poorly studied. In the aspect of elaboration of the optimization model of information economy it is necessary to reconsider the existing tools of tax administration and supplement them for increasing the efficiency and security of implementation of measures for formation of the revenue part of the budget. These issues are studied in this chapter. The research is performed with the help of the methods of graphical presentation of information, statistical analysis, trend analysis, method of comparison, analogy, and systematization.

3   Results

Evaluation of efficiency of the implemented tax policy showed that dynamics of collected taxes is positive, with growth of the volumes of paid taxes and fees—which shows efficiency of the fiscal policy (Table 1). However, during the whole analyzed period the budget is deficit, which is caused by growth of expenditures in 2017 by 62%, as compared to 2010. The increased volumes of collected taxes do not compensate for reduction of oil and gas revenues, which dropped after the 2014 crisis. A vivid growth is seen for the following taxes: VAT (internal) with growth of 131%, excise duties—698%, excise duties for imported goods—159%, personal income tax—198%.

Non-oil and gas revenues of the state budget grew in 2017 by 103%, as compared to 2010; growth of oil and gas revenues constituted 54%. Apart from revenues, the total sum of expenditures grew by 62% in 2017, which led to increase of deficit by 73%. The existing picture of formation of the revenue part of the budget shows high potential of non-oil and gas revenues, which could become additional sources of financing of the budget's expenditures. Taxes have a large part in replenishment of the revenue part of the budget.

The received deficit of revenues increases risks of social and economic security of the state, as certain socio-economic obligations cannot be performed due to lack of state financing. That's why tax security of the state and search for sources of its increase could become a perspective direction for future state development.

In this case, it is necessary to speak of quality of the existing of tax system. For evaluating the quality of the tax system, it is necessary to consider the ranking of different countries, compiled by international agencies. In particular, the ranking

Table 1 Statistics of revenues and expenditures of the budget in 2010–2017, RUB billion

	2010	2011	2012	2013	2014	2015	2016	2017
Revenues, total	8305.4	11,367.7	12,855.5	13,019.9	14,496.9	13,659.2	13,460.0	15,088.9
Oil and gas revenues	3830.7	5641.8	6453.2	6534.0	7433.8	5862.7	4844.0	5971.9
Non oil and gas revenues, including	4474.7	5725.9	6402.4	6485.9	7063.1	7796.6	8616.0	9117.0
VAT (internal)	1328.7	1753.2	1886.1	1868.2	2181.4	2448.3	2657.4	3069.9
Excise duties	113.9	231.8	341.9	461.0	520.8	527.9	632.2	909.6
Personal income tax	255.0	342.6	375.8	352.2	411.3	491.4	491.0	762.4
VAT on imported goods	1169.5	1497.2	1659.7	1670.8	1750.2	1785.2	1913.6	2067.2
Excise duties on imported tax	30.1	46.6	53.4	63.4	71.6	54.0	62.1	78.2
Import duties	587.5	692.9	732.8	683.8	652.5	565.2	563.9	583.2
Expenditures, total	10,117.5	10,925.6	12,895.0	13,342.9	14,831.6	15,620.3	16,416.4	16,420.
Deficit	−1812.0	442.0	−39.4	−323.0	−334.7	−1961.0	−2956.4	−1331.4

Source Compiled based on the data "Annual information on execution of the federal budget". Access: https://www.minfin.ru/ru/statistics/fedbud/execute/ (Accessed: 10.07.2018)

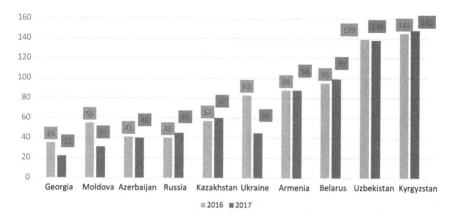

Fig. 1 Ranking of tax load of the CIS countries "Paying Taxes 2017". *Source* Paying Taxes 2017: the key countries of the region lose their positions. Access: http://rescue.org.ru/ru/news/analytics/ 5549-paying-taxes-2017-klyuchevye-strany-regiona-sdayut-pozitsii (Accessed: 10.07.2018)

Paying Taxes 2017, compiled by the World Bank and Pricewaterhouse Coopers, could be used. This ranking is peculiar for the fact that evaluation is performed according to several criteria: administration and payment of taxes; processes in the sphere of taxation; inspections; budget compensations. Let us consider the results of the ranking Paying Taxes 2017 (Fig. 1).

The ranking showed that the main tendency is aimed at reduction of time for companies' executing the tax law.

Recently, integral evaluation was supplemented by the indicator Index "after payment". This criterion reflects the aggregate tax load that is put on the company after payment of taxes (1–100 points). According to this indicator, the level of effectiveness of tax systems is determined by formation of motivating environment for payment of taxes. The main qualities of such systems are transparency, convenience, and speed of administration. That's why for solving the tasks of preservation of security of taxpayers' economic activities and preventing the risks of state's shortfall of tax revenues it is necessary to build the systems of secure tax administration, based on application and implementation of new information technologies—i.e., increase of system's tax security.

The state strived for preservation of integrity and effectiveness of the tax system, which ensures gradual transition to the information platforms of implementation of functions of tax administration. For this purpose, the Federal Tax Service of the RF adopts annual acts on the Plan of informatization. For example, there was a Decree dated March 30, 2018, №. 74 "The plan of informatization for the next financial year (2018) and the planned period of 2019 and 2020". Let us view the main tools of digital technologies that are given in this document (Table 2).

Table 2 shows certain information tools that are implemented in the work of the Federal Tax Service of the RF. The set target indicators of implementation of certain measures of informatization are rather debatable. Thus, the measures

Table 2 Certain tools of digital technologies that are used and implemented into the work of the Federal Tax Service of the RF

Indicator	Financing from the budget, RUB thousand			Target indicator			
	Plan year	1 year	2 years	Title	Plan year	1 year	2 years
Development of AIS "Nalog-3"	60,690	50,051.60	0	Number of state services that are transferred into the electronic form according to the set requirements and that are provided with the help of the information system	5	5	5
				Number of state functions that are performed by the state body with the help of the information system	15	15	15
				Measures for protection of information according to the requirements	Yes	Yes	Yes
Exploitation of AIS "Nalog-3"	792,903.40	764,037	764,037	Measures for protection of information according to the requirements	Yes	Yes	Yes
Exploitation of AIS "Markirovka"	35,380.10	30,343	30,343	Measures for protection of information according to the requirements	Yes	Yes	Yes
Exploitation Information resource of accounting and statistical reports of Russian and foreign organizations	7941.70	7941.70	7941.70	–	–	–	–
Exploitation Information resource of evaluation and analysis of commodity market	14,756	14,756	14,756	–	–	–	–

(continued)

Table 2 (continued)

Indicator	Financing from the budget, RUB thousand			Target indicator			
	Plan year	1 year	2 years	Title	Plan year	1 year	2 years
Exploitation Telecommunication infrastructure if "Nalog-Servis" of the Federal Tax Service of the RF which provides external communications	134,215.80	91,368	91,368	–	–	–	–
…	…	…	…	…	…	…	…
Total	2,754,488.60	2,194,071.40	2,193,829				

Source Plan of informatization of the Federal Tax Service. Access: https://www.nalog.ru/rn77/about_fts/fts/activities_fts/ (Accessed 09.06.2018)

"Exploitation of the AIS 'Nalog-3'" and "Exploitation of the AIS 'Markirovka'" contain such indicators as measures on protection of information according to requirements. We think that such indicators do not provide any information and do not reflect peculiarities of these measures. As a matter of fact, the indicator is nominal, without any rational component.

Besides, it is unclear why the Plan of transfer does not include measures for development and exploitation of the systems "NDS-2", KKT, and "EGR ZAGS", which are constantly improved by the tax service. These systems have been working for several years, but their operation is not envisaged by the law completely. That's why it is necessary to continue development and improvement of these systems, as their implementation is not always affordable for business and leads to its bankruptcy. In this case, it is necessary to expand possibilities of these platforms by integration with the services of the Federal Customs Service of the RF and federal law enforcement for the necessary information on a business partner and his portfolio to become accessible for all subjects of economic relations.

Such work is conducted by tax bodies with the help of the Federal Government Information System, which is a data base on all citizens. Thus, there is experience of creation of one platform for individuals, so it is necessary to create a similar platform for all economic subjects that conduct entrepreneurial activities on the territory of Russia. The volumes of financing of these tools are associated with the state policy of creation of the model of information economy, which is aimed at increase of quality by implementing information and telecommunication technologies in all spheres of society's life (Table 3).

Thus, according to the data of Table 1, we see that the level of society's informatization is evaluated according to 6 main indicators, which dynamics is planned until 2020. According to the main target indicators of formation of information society, the Federal Tax Service of the RF implements information technologies on the basis of artificial intelligence for all directions of tax administration.

For strengthening the national tax security and supporting the programs of import substitution, the Road map for provision of Strategy of the Federal Tax Service is used, which was approved by the Decree of the Russia's FTS dated January 18, 2018, No. MMB-7-6/24@ "Regarding adoption of the Strategy of the Russia's FTS for import substitution of IT infrastructure and software that are applied in automatized information systems of the Russia's FTS in view of transition to dominating usage of products and solutions of domestic manufacturers", which means full transition of all online services to domestic software. The main stages of this transition are given in Fig. 2. It is seen that the decision on development of the Strategy of the Russia's FTS was made in 2016. This strategy is aimed at development of the domestic market of IT. Evaluation of plan-graph of transition to domestic software showed that the main services in the Decree are text and table editors, presentations, office packages, operational systems, mail software, search systems, systems of electronic document turnover, anti-virus programs, multimedia, and internet browsers. According to the plan of transition for 2018–2020, these services should be based on domestic software by 50–8% by 2018, and by 2020 the transition should constitute 80–100%.

Table 3 The main target indicators of the state program of the RF "Information society (2011–2020)"

Indicator	Measuring unit	2010	2011	2012	2013	2014	2015	2016	2017	2018	2019	2020
Russia's position in the international ranking according to the index of development of information technologies	–	48	48	48	Among top-40 countries	Among top-40 countries	Among top-20 countries	Among top-10 countries	Among top-10 countries	Among top-10 countries	Among top-10 countries	Among top-10 countries
Share of people that use electronic state services	%	11	20	25	30	35	40	50	60	70	70	70
Share of population that do not use the Internet in the total number of population	%	23	21	19	16	11	7	5	5	4	3	2
Level of differentiation of subjects of the RF according to integral indicators of information development	Units	–	3.6	2.9	2.6	2.3	2	2	1.9	1.9	1.8	1.8
Share of households with access to the Internet from a home computer in the total number of households	%	41.3	50.2	55.1	63.4	66.8	69.9	72.8	75.4	77.9	80.3	82.6
Number of highly-efficient jobs for the type of economic activities "communications"	Thousand units	–	–	358.2	339.7	361.5	381.5	401.5	421.5	441.5	461.5	483.0

Source Compiled based on the data of the National Program of RF "Information society (2011–2020)". Adopted by the Decree of the Government of the RF dated April 15, 2014 No. 313. SPS Consultant Plus. Access: http://www.consultant.ru/document/cons_doc_LAW_162184/ (Accessed: 03.06.2018)

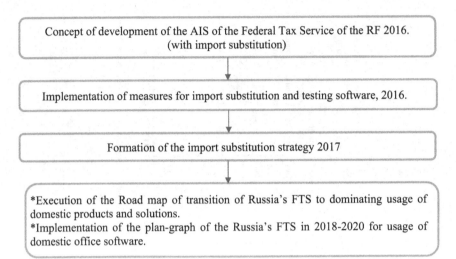

Fig. 2 Stages of structural approach to import substitution in the work of Russia's FTS. *Source* Information technologies in the Federal Tax Service. Access: http://www.tadviser.ru/index.php/ (Accessed: 10.06.2018)

Total number of services on the official web-site of the Russia's FTS is 50. The structure and popularity of services as of 2017 are shown in Fig. 3.

Figure 3 shows that the Russia's FTS tries to develop not only the fiscal function (function of collection of taxes) but also the service function for taxpayers and other bodies. Development of the service online system of work of the tax service is aimed at optimization of spent time and finances for the budget and for citizens. The tax service continues implementing new devices, developing mobile apps, and improving online accounts for taxpayers. All these measures allow for remote work and observation of strict formalization in communication between tax bodies and citizens.

Concluding the analysis of the existing tax system of Russia and its work for provision of tax security, it is possible to note that the main work has already been done, the first results have been received, and further exploitation of services is expected. The main stage for provision of Russia's tax security is development and implementation of the Strategy of the Russia's Federal Tax Service for import substitution of IT infrastructure and software, applied in automated information systems of the Russia's Federal Tax Service. This document is to ensure the development of domestic manufacturers of information technologies, reduce unsanctioned accesses into the Russian systems and data base, create conditions for continuous work of the AIS of the Russia's Federal Tax Service, update software, and reduce dependence on manufacturers and conditions of international trade.

Fig. 3 The main services of the Federal Tax Service of the RF and quality of their usage for 2017. *Source* Information technologies in the Federal Tax Service. Access: http://www.tadviser.ru/index. php/ (Accessed: 10.06.2018)

4   Conclusion

Thus, the performed evaluation of the level of tax security in the conditions of information economy showed that the Russia's Federal Tax Service aims at stage-by-stage implementation of information and digital services into their practical activities. For supporting the import substitution programs, the Strategy of transfer of all information services to domestic software is created. All measures that are implemented by the Russia's Federal Tax Service correlate with the main target indicators of the National Program "Information society (2011–2020)". Implementation of the planned measures by the Russia's Federal Tax Service will allow reducing political risks in the conditions of economic sanctions and developing domestic production, thus increasing taxation base and operative cooperation with developers of software. Eventually, these measures will allow building own effective and secure system of taxation, which is protected from external threats.

References

Aristovnik, A., & Obadić, A. (2015). The impact and efficiency of public administration excellence on fostering SMEs in EU countries. *Amfiteatru Economic, 17*(39), 761–774.

Bondarenko, Y. G. (2015). Public administration efficiency increase in investment management. *Actual Problems of Economics, 172*(10), 89–94.

Comerio, M., Batini, C. (2016). Efficiency vs efficacy driven service portfolio management in a public administration (invited paper). In *Proceedings–2015 IEEE 8th International Conference on Service-Oriented Computing and Applications, SOCA 2015*, 7399103 (pp. 139–146).

Dai, X. (2013). *Politics of digital development: Informatization and governance in China. Digital world: Connectivity, creativity and rights* (pp. 34–51). London: Taylor and Francis.

Hawes, C., & Li, G. (2017). Transparency and opaqueness in the Chinese ICT sector: A critique of Chinese and international corporate governance norms. *Asian Journal of Comparative Law, 12* (1), 41–80.

Huggins, C., & Frosina, N. (2017). ICT-driven projects for land governance in Kenya: Disruption and e-government frameworks. *GeoJournal, 82*(4), 643–663.

Méndez Reátegui, R., Coca Chanalata, D. G., & Alosilla Díaz, R. (2016). The efficiency of public administration: A compartive analysis of the peruvian and ecuadorian cases from a neo institutional approach|[La eficiencia de la administración: Un análisis comparado desde el enfoque neo institucional de los casos peruano y ecuatoriano]. *Revista General de Derecho Administrativo, 2016*(43), 12–19.

Monitoring the development of the information society in the Russian Federation. Available at: http://www.gks.ru/wps/wcm/connect/rosstat_main/rosstat/ru/statistics/science_and_innovations/it_technology. Accessed March 3, 2018.

Nasir, A., Shahzad, M., Anwar, S., Rashid, S. (2017). Digital governance: Improving solid waste management through ICT reform in Punjab. In *ACM International Conference Proceeding Series*, Part F132087, 3136600.

Pessoa, A. A. M., Justino, A. N. P., De Farias, F. H. C., Lima, P. T. D., De Sousa, V. R. M., et al. (2016). Analysis of the efficiency, efficacy and effectiveness in public administration: The case of IDEMA/RN|[Análise da eficiência, eficácia e efetividade na administração pública: O caso do IDEMA/RN]. *Espacios, 37*(8), c. 8.

Plotnokov, V., Fedotova, G. V., Popkova, E. G., & Kastyrina, A. A. (2015). Harmonization of strategic planning indicators of territories' socioeconomic growth. *Regional and Sectoral Economic Studies, 15*(2), 105–114.

Romanova, T. F., Andreeva, O. V., Meliksetyan, S. N., & Otrishko, M. O. (2017). Increasing of cost efficiency as a trend of public law entities' activity intensification in a public administration sector. *European Research Studies Journal, 20*(1), 155–161.

Sakil, A. H. (2017). ICT, youth and urban governance in developing countries: Bangladesh perspective. *International Journal of Adolescence and Youth*, 1–16.

Sukhodolov, A. P., Popkova, E. G., & Kuzlaeva, I. M. (2018). Methodological aspects of study of internet economy. *Studies in Computational Intelligence, 714*, 53–61.

Vertakova, Y., Plotnikov, V., & Fedotova, G. (2016). The system of indicators for indicative management of a region and its clusters. *Procedia Economics and Finance, 39*, 184–191.

Fedotova, G. V., Kulikova, N. N., Kurbanov, A. K., Gontar, A. A.: Threats to food security of the Russia's population in the conditions of transition to digital economy [e-source]. In E. G. Popkova (Ed.), *The impact of information on modern humans* (pp. 542–548). Switzerland: Springer International Publishing AG (part of Springer Nature) (2018). https://link.springer.com/content/pdf/bfm%3A978-3-319-75383-6%2F1.pdf.

New Types of Taxes and Forms of Taxation in the Conditions of Information Economy: Perspectives of Optimization

Olga V. Titova, Elena I. Kostyukova,
Natalia M. Boboshko and Irina P. Drachena

Abstract *Purpose* The purpose of the work is to determine new types of taxes and new forms of taxation in the conditions of information economy and to determine the perspectives of their optimization. *Methodology* The methods of systemic and problem analysis, logical analysis (analysis of causal connections), and method of formalization (graphic presentation of information) are used. *Results* It is determined that in the conditions of information economy all three closely interconnected tendencies are manifested: growth of the value of information and its becoming the key resource, distribution of information technologies and their becoming a special type of business assets, and formation and development of Internet entrepreneurship. Under the influence of these tendencies, which are already seen in economic systems, possibilities of collection of traditional types of taxes (with the help of traditional forms of taxation, based on corporate tax reports) will be limited. This will lead to establishment of a new balance in the tax system, at which traditional types of taxes will be replaced with new types of taxes, collected with the help of the corresponding new forms of taxation. These are tax on information, collected in the Internet form, on information technologies, collected in the ICT form, and tax on Internet sales, collected in the electronic form. These new types of taxes and new forms of taxation, which will be applied in the

O. V. Titova (✉)
Altai State University, Barnaul, Russia
e-mail: otitova82@icloud.com

E. I. Kostyukova
Stavropol State Agrarian University, Stavropol, Russia
e-mail: Elena-kostyukova@yandex.ru

N. M. Boboshko
Moscow University of the Ministry of Internal Affairs of the Russian
Federation Named After V.J. Kikot, Moscow, Russia
e-mail: Natmb@iist.ru

I. P. Drachena
State Budgetary Educational Institution of Higher Education of the
Moscow Region, Technological University, Moscow Region, Korolev, Russia
e-mail: Irina.drachena@mail.ru

© Springer Nature Switzerland AG 2019
I. V. Gashenko et al. (eds.), *Optimization of the Taxation System: Preconditions,
Tendencies, and Perspectives*, Studies in Systems, Decision and Control 182,
https://doi.org/10.1007/978-3-030-01514-5_26

conditions of information economy, will influence entrepreneurial structure and individuals. *Recommendations* For preventing a social crisis due to coming transformations of the tax system in the conditions of information economy and for protecting the interests of the state in the aspect of full-scale collection of taxes, the authors offer recommendations on optimization of new types of taxes and new forms of taxation, related to establishment of lower tax rates.

Keywords New types of taxes · New forms of taxation · Tax optimization Information economy

JEL Classification E62 · H20 · K34

1   Introduction

Information economy is a new type of socio-economic system that is peculiar for wide informatization of economic phenomena and processes. At present, increase of the scale and acceleration of the process of informatization in a lot of spheres of national economy lead to signs of transition to information economy. An example could be the sphere of communications, in which telecommunications, the Internet, and new information technologies oust other means of communications (radio, telephone, etc.). Establishment of information economy is contradictory from the point of view of taxation.

On the one hand, information economy opens new opportunities for increase of convenience of taxation of economic subjects due to development of the system of E-government, which allows for automatic receipt and sorting of information on individual and corporate taxation and changes in the tax system, as well as for remote payment of taxes. For the state, this means reduction of expenditures for tax administration and control and growth of collection of taxes due to automatization of taxation.

On the other hand, information economy could become a reason for social crises, as in its striving for maximum transparency and manageability, the state tests innovations in taxation, accessible due to informatization of economy, and then ousts usual traditional, without providing taxpayers with sufficient time for adaptation to these changes. For example, there are active attempts by the state to oust cash, replacing it with electronic payment means.

This could increase collection of corporate tax, added value tax, personal income tax, etc., but this will probably cause public opposition and protests—especially from the categories of the population that are least inclined to innovations (e.g., pensioners). This actualizes the problem of adaptation of the taxation system to the conditions of information economy and optimization of this system.

The working hypothesis of the research is that in the conditions of information economy there appear new types of taxes and new forms of taxation. For preparing interested parties for their emergence and implementation into the practice of taxation, and for maximum usage the possibilities of development of the tax system that open in the conditions of information economy, it is necessary to determine new types of taxes and new types of taxation in the conditions of information economy and to determine perspectives of their optimization, which is the purpose of this work.

2 Materials and Method

The foundations and practical experience of formation of information economy in the modern economic systems are discussed in the works: Bogoviz et al. (2017), Gladilin et al. (2019), Protopopova et al. (2019), Sukhodolov et al. (2018a, b), Zaytsev et al. (2019), and Zmiyak et al. (2019).

Possibilities, perspectives, and consequences of informatization of tax system are studied in the works: Alkan and Surmeneli (2018), Alm et al. (2017), Becchetti et al. (2017), Boudreau and MacKenzie (2018), Chen et al. (2018), Chen and Lin (2017), Gashenko et al. (2018), Kolstad and Wiig (2018), and Passalacqua et al. (2018).

The performed literature overview showed that the existing studies and publications focus on general issues of formation of information economy and informatization of tax system. Detalization of these processes, related to determining and optimizing new types of taxes and new forms of taxation that appear in the conditions of information economy, requires additional research which is done in this work. We use the methods of systemic and problem analysis, logical analysis (analysis of causal connections), and formalization (graphic presentation of information).

3 Results

As a result of studying the process of formation of information economy, we determined the following tendencies of this process and the emerging new types of taxes and forms of taxation (Table 1).

Lew us study the tendencies of information economy and emerging new types of taxes and forms of taxation in detail and determine the perspectives of their optimization. The first tendency is related to growth of the value of information and its becoming a key resource. In the aspect of taxation, this tendency will lead to emergence of new methods of creation of added value, based on information, and increase of the chains of added value by means of including the processes of creation, distribution, and usage of information.

This allows implementing tax on information, which could be an independent tax or an addition to added value tax. At that, the objects of taxation will be the following:

Table 1 Tendencies of information economy and emerging new types of taxes and forms of taxation

Tendencies of information economy	New types of taxes	New forms of taxation
Growth of the value of information and its becoming the key resource	Tax on information	Internet form
Distribution of information technologies and their becoming a special type of business assets	Tax on information technologies	ICT[a] form
Formation and development of internet entrepreneurship	Tax on Internet sales	Electronic form

Source Compiled by the authors
[a]Information and communication technologies

- innovations (new information), used in the production process or created for selling;
- process of creation of new information (R&D) and its bearers (company workers);
- process of information exchange in the interests of business (e.g., promotion of products in social networks).

As information is the moving force of development of information economy, tax on information should not restrain the processes of its creation, dissemination, and usage, but quite on the contrary—it has to stimulate activity in them. That's why optimization of tax on information envisages establishment of lower tax rates as compared to standard rates. For example, in modern Russia added value tax is 18%, and successful optimization of information tax envisages application at the stages of the chain of added values, at which information is created, disseminated, and used, of reduced tax rate (10%).

At that, the Internet form of taxation becomes available, within which the tax service uses the Internet to collect from the interested parties feedback on the issues of execution of economic operations and their taxation. Economic subjects, being interested in access to reduced rates of added value tax (or information tax), will inform the tax service on execution of economic operations in the process of which information is created, disseminates, and used.

This will ensure successful collection of information tax, at which a reason for taxation is not the fact of selling the product (as with added value tax), but the very fact of execution of economic operations that precede selling of products. Optimization of this form of taxation is related to creation of clear and vivid terms of payment of information tax and development (in the aspect of convenience, continuous work, and safety) of the system of E-government, through which Internet communication feedback will be collected.

The second tendency is dissemination of information technologies and their becoming a special type of business assets. Development of Internet entrepreneurship in the conditions of information economy will inevitably lead to reduction of the fixed assets that are used in business and, therefore, reduction of tax potential of corporate tax on property. For preserving the volume of revenues into the state budget, it should be replaced by tax on information devices and technologies. At that, the objects of taxation are as follows:

- computer devices that are used in entrepreneurial activities;
- software that is used in entrepreneurial activities;
- mobile communications that are used in entrepreneurial activities;
- access to the Internet that is used in entrepreneurial activities.

With the traditional method of conduct of entrepreneurial activities in the conditions of the general regime of taxation, computer devices is put on the balance of the company and is imposed with property tax. However, Internet companies will probably fall under a special tax regime, at which no corporate tax accounting is envisaged. That's why there will be a new (optimal in this case)—ICT form of taxation, at which information devices and technologies will be equipped with the Internet of Things, through which they will automatically inform the tax service on their usage in entrepreneurial activities, based on which tax on information devices and technologies will be accrued and paid automatically.

The third tendency is formation and development of Internet entrepreneurship. This will probably lead to reduction of business activity of traditional entrepreneurship, so in order to prevent the impossibility of collection of corporate tax in the full scale, the tax on Internet sales will be implemented. The objects of taxation here will be as follows:

- traditional goods, which have material form and are sold via the Internet (e.g., clothes, household appliances, etc.);
- electronic goods, which do not have material form and which are created and sold primarily on the Internet (e.g., goods that are sold in online games).

Within the studied concept, there also emerges an electronic form of taxation, which envisages tax on conducted electronic payments by economic subjects. As traditional tax administration and control over Internet sales are difficult, there's a probability that all electronic payments and transfers of economic subjects (excluding close relatives and special cases) will be considered payment for Internet sales (both in the B2C and C2C forms).

Based on this, tax with a fixed rate will be paid from all electronic payments and transfers. Optimization of this type of tax and this form of taxation envisages establishment of a lower rate that with corporate tax, as the object of taxation is not profit, but income. For example, in modern Russia the rate of corporate tax is 20%, and the rate of personal income tax—13%. To avoid public protests, related to complexity of proving whether a specific electronic payment or transfer is income, the rate of tax on Internet sales is supposed to constitute 3%.

4 Conclusion

Thus, in the course of the research the offered hypothesis is proved; it is determined that in the conditions of information economy three closely interconnected tendencies are manifested: growth of the value of information and its becoming the key resource, dissemination of information technologies and their becoming a special type of business assets, and formation and development of Internet entrepreneurship. Under the influence of these tendencies, which are already seen in the modern economic systems, possibilities of collection of traditional types of taxes (with the help of traditional forms of taxation, based on corporate tax accounting) will be limited.

This will lead to formation of new balance in the tax system, at which traditional types of taxes will be replaced by new types of taxes, which are collected with the corresponding new forms of taxation. These are tax on information that is collected in the Internet form, tax on information technologies that is collected in the ICT form, and tax on Internet sales that is collected in the electronic form. These new forms of taxes and forms of taxation, which will be applied in the conditions of information economy, will be applied to entrepreneurial structures and individuals.

For preventing a social crisis due to coming transformations of the tax system in the conditions of information economy and for protecting the state's interests in the aspect of full-scale collection of taxes, the authors offer recommendations for optimization of new types of taxes and new forms of taxation, related to establishment of lower tax rates. As expected, this will allow gaining advantages from the specific "scale effect", at which all economic operations will fall under taxation due to increase of transparency and controllability of economic activities, and low tax rates will make taxation profitable for economic subjects and for the state.

References

Alkan, M., & Surmeneli, H. G. (2018). Development of an advertisement tax system based on a geographic information system. *Proceedings of the Institution of Civil Engineers: Municipal Engineer, 171*(2), 93–104.

Alm, J., Bloomquist, K. M., & McKee, M. (2017). When you know your neighbour pays taxes: Information, peer effects and tax compliance. *Fiscal Studies, 38*(4), 587–613.

Becchetti, L., Pelligra, V., & Reggiani, T. (2017). Information, belief elicitation and threshold effects in the 5X1000 tax scheme: A framed field experiment. *International Tax and Public Finance, 24*(6), 1026–1049.

Bogoviz, A. V., Ragulina, Y. V., Komarova, A. V., Bolotin, A. V., & Lobova, S. V. (2017). Modernization of the approach to usage of region's budget resources in the conditions of information economy development. *European Research Studies Journal, 20*(3), 570.

Boudreau, C., & MacKenzie, S. A. (2018). Wanting what is fair: How party cues and information about income inequality affect public support for taxes. *Journal of Politics, 80*(2), 367–381.

Chen, C.-W., Hepfer, B. F., Quinn, P. J., & Wilson, R. J. (2018). The effect of tax-motivated income shifting on information asymmetry. *Review of Accounting Studies, 2*(1), 1–47.

Chen, T., & Lin, C. (2017). Does information asymmetry affect corporate tax aggressiveness? *Journal of Financial and Quantitative Analysis, 52*(5), 2053–2081.

Gashenko, I. V., Zima, Y. S., Stroiteleva, V. A., & Shiryaeva, N. M. (2018). The mechanism of optimization of the tax administration system with the help of the new information and communication technologies. *Advances in Intelligent Systems and Computing, 622,* 291–297.

Gladilin, A. V., Dotdueva, Z. S., Klimovskikh, Y. A., Labovskaya, Y. V., & Sharunova, E. V. (2019). Establishment of information economy under the influence of scientific and technical progress. *Advances in Intelligent Systems and Computing, 726,* 46–55.

Kolstad, I., & Wiig, A. (2018). How does information about elite tax evasion affect political participation: Experimental evidence from Tanzania. *Journal of Development Studies, 2*(1), 1–18.

Passalacqua, A. B. S., Mazz, A., Pistone, P., Quiñones, N., Roeleveld, J., Schoueri, L. E., et al. (2018). Tax information exchange agreements and the prohibition of retroactivity. *Intertax, 46*(5), 368–389.

Protopopova, N. I., Grigoriev, V. D., & Perevozchikov, S. Y. (2019). Information and digital economy as an economic category. *Advances in Intelligent Systems and Computing, 726,* 300–307.

Sukhodolov, A. P., Popkova, E. G., & Kuzlaeva, I. M. (2018a). Peculiarities of formation and development of internet economy in Russia. *Studies in Computational Intelligence, 714,* 63–70.

Sukhodolov, A. P., Popkova, E. G., & Kuzlaeva, I. M. (2018b). Methodological aspects of study of internet economy. *Studies in Computational Intelligence, 714,* 53–61.

Zaytsev, A. G., Plakhova, L. V., Legostaeva, S. A., Zakharkina, N. V., & Zviagintceva, Y. A. (2019). Establishment of information economy under the influence of scientific and technical progress: New challenges and possibilities. *Advances in Intelligent Systems and Computing, 726,* 3–10.

Zmiyak, S. S., Ugnich, E. A., & Krasnokutskiy, P. A. (2019). Generation and commercialization of knowledge in the innovational ecosystem of regional university in the conditions of information economy establishment in Russia. *Advances in Intelligent Systems and Computing, 726,* 23–31.

Conclusions

The 21st century is a period of domination of digital technologies, which allow optimizing various components of economic activities. The taxation system cannot ignore the global tendency of digitization and should become one of the key objects of reformation in the process of formation of the digital economy. Optimization of the taxation system requires mandatory consideration and profitable use of the possibilities of new information and communication technologies.

The given substantiation of the necessity for optimization of the modern Russia's taxation system allowed actualizing the problem of increasing its effectiveness. Determining the key directions of taxation optimization in modern Russia provided a basis for correction of state tax policy in view of actual needs of economy. Offering new and perspective ways, methods, and mechanisms of taxation optimization in modern Russia will allow increasing its effectiveness by preventing tax evasion and by achieving favorable influence of the taxation system on small and medium business.

According to the results of a complex of research, presented in this book, informatization (digitization) has a large potential in the sphere of the modern Russia's taxation optimization, allowing solving the problems caused by imperfection of state management of the taxation system, irrational spending of state budgets' assets, tax opportunism, "free rider problem" in taxes, and low level of involvement of interested parties in the process of development and implementation of state's tax policy.

Moreover, digital modernization of the modern Russia's taxation system will allow overcoming its current crisis, which is manifested in critical reduction of its effectiveness due to increase of deficit of state budgets of all levels of the budget system and non-budget funds and large and increasing volume of shadow economy. Informatization (digitization) ensures reduction of the risk component of functioning and development of the taxation system due to limitation of influence of "human factor" on it, by intellectual support for organizational and managerial decisions.

© Springer Nature Switzerland AG 2019
I. V. Gashenko et al. (eds.), *Optimization of the Taxation System: Preconditions, Tendencies, and Perspectives*, Studies in Systems, Decision and Control 182, https://doi.org/10.1007/978-3-030-01514-5

Advantages of digital modernization of the system of taxation include also growth of its transparency, openness, and controllability. The obtained results showed that it is possible to achieve synergetic effect by supplementing application of digital technologies for increasing effectiveness of the taxation system by creation of tax stimuli for digitization of economy. This will allow harmonizing optimization of the taxation system and general progress of formation and development of the digital economy.

At the same time, a lot of issues of optimization of the taxation system remained open and were supplemented by new issues that emerged as a result of performed results. In particular, the issue of provision of reliability and security of digital taxation, social adaptation to conditions of digital modernization of taxation system, and development of the system of education and training of personnel for digital taxation became topical. These issues should be studied during further research.

Printed in the United States
By Bookmasters